ROUTLEDGE LIBRARY EDI
LIBRARY AND INFORMATION SCIENCE

Volume 11

BIOGRAPHIES OF SCIENTISTS FOR SCI-TECH LIBRARIES

BIOGRAPHIES OF SCIENTISTS FOR SCI-TECH LIBRARIES
Adding Faces to the Facts

Edited by
TONY STANKUS

R Routledge
Taylor & Francis Group

LONDON AND NEW YORK

First published in 1991 by The Haworth Press, Inc.

This edition first published in 2020
by Routledge
2 Park Square, Milton Park, Abingdon, Oxon OX14 4RN

and by Routledge
52 Vanderbilt Avenue, New York, NY 10017

Routledge is an imprint of the Taylor & Francis Group, an informa business

© 1991 The Haworth Press, Inc.

British Library Cataloguing in Publication Data
A catalogue record for this book is available from the British Library

ISBN: 978-0-367-34616-4 (Set)
ISBN: 978-0-429-34352-0 (Set) (ebk)
ISBN: 978-0-367-43351-2 (Volume 11) (hbk)
ISBN: 978-0-367-43388-8 (Volume 11) (pbk)
ISBN: 978-1-00-300285-7 (Volume 11) (ebk)

Publisher's Note
The publisher has gone to great lengths to ensure the quality of this reprint but points out that some imperfections in the original copies may be apparent.

Disclaimer
The publisher has made every effort to trace copyright holders and would welcome correspondence from those they have been unable to trace.

Biographies of Scientists for Sci-Tech Libraries: Adding Faces to the Facts

Tony Stankus
Editor

The Haworth Press, Inc.
New York • London • Sydney

Biographies of Scientists for Sci-Tech Libraries: Adding Faces to the Facts has also been published as *Science & Technology Libraries*, Volume 11, Number 4, 1991.

The Haworth Press, Inc., 10 Alice Street, Binghamton, NY 13904-1580
EUROSPAN/Haworth, 3 Henrietta Street, London WC2E 8LU England
ASTAM/Haworth, 162-168 Parramatta Road, Stanmore, Sydney, N.S.W. 2048 Australia

Library of Congress Cataloging-in-Publication Data

Biographies of scientists for sci-tech libraries : adding faces to the facts / Tony Stankus, editor.
    p. cm.
    "Has also been published as Science & technology libraries, volume 11, number 4, 1991" — T.p. verso
    Includes bibliographical references.
    ISBN 1-56024-214-0 (acid free paper)
    1. Scientists — Biography. 2. Engineers — Biography. 3. Science — Bio-bibliography. 4. Engineering — Bio-bibliography. I. Stankus, Tony.
Q141.B535 1991
509.2 — dc20
[B]   91-31241
                                             CIP

Biographies of Scientists for Sci-Tech Libraries: Adding Faces to the Facts

CONTENTS

Preface

Biographies of Scientists for Sci-Tech Libraries: Adding Faces to the Facts identifies biographies of scientists from a selection of scientific fields. It differs from most compilations of biographical works in that its primary purpose is to identify titles that provide insights into the variety and drama of the lives of these men and women beyond their more well-known contributions and Nobel Prize accomplishments.

As author, Dena Rae Thomas ("Plant Breeders and Plant Genetics") explains in her introduction, "fewer students in the U.S. are choosing to pursue graduate studies or careers in science. It seems that some exploration of scientists' lives might inspire individual students to follow any leanings toward science they may have. The biographies . . . described here were chosen because they attempt to show the reader the whole personality of the biographee including contributions made to his/her field."

As Editor, Tony Stankus, suggests in his introduction, these biographies may also serve to "shatter the stereotypes" that both librarians and readers too often carry with regard to scientists and their lives — as "monomaniacal and uninteresting" people.

This volume is the inspiration of Tony Stankus. Tony's own contribution clearly illustrates the humor and humanity associated with the work of scientists, providing insights into their personalities that might otherwise not be expected. I am extremely grateful to him for developing and compiling this volume. Together we hope these annotated listings will be helpful to sci-tech librarians in furthering their understanding and appreciation of science as a broadly-based and creative experience — and to use these titles to share this understanding with students and other readers.

As Editor of the "Special Collections" section, Tony has also arranged with Julie Hurd and Gladys Odegaard (University of Illinois, Chicago) for a special offering on the science of spectroscopy and the literature of spectra. Sections describing "New Reference Books in Sci-Tech" and "Sci-Tech Online" are also included.

Cynthia A. Steinke
Editor
Sciences & Technology Libraries

Introduction

People grow up knowing what a doctor, priest, mail carrier, or firefighter is and does, but rarely have a feel for who scientists are, and what scientists do. During the formative years of many readers, school and small town public librarians may not always be good sources of this information for at least two practical reasons. The first is that those librarians must buy what seems likely to circulate most readily. The biography of an historical figure or that of a star athlete may see more action in these collections than the life of a chemist. Second, most librarians have not been drawn from the ranks of the scientifically eager. They have had little contact beyond a few required courses during their undergraduate years with the people who have made up science. Librarians are not likely to have much anecdotal background to start up and fuel the young reader's interest.

Even university majors in science and engineering are relatively underexposed to role models in the literature of science. The pressure to present facts, concepts, and calculation pathways in a limited amount of time and space has driven a good deal of biographical detail out of most university lectures, texts and assigned reading. This is particularly unfortunate, for future scientists can find encouragement, inspiration, and an understanding of how their specialty came to be through a look at the great lives in science. Specialties did not just happen. People seized on a different theme or idea, and had some early successes with their novelty. From that success came followers. When the followers became numerous enough, they identified more closely with each other than with the fields for which they originally trained.

Librarians should be required to read some of these biographies for the sake of seeing how new journals came into being. Noting the number of fields which developed in Europe, Japan, and even within the Third World might dispel at least some notions that

North Americans are responsible for all successes in scientific specialties or that U.S. libraries could entirely forgo foreign publications and remain largely in the know. Librarians could also use some shattering of the stereotypes that so many of them carry with regard to scientists. These stereotypes concern the purportedly monomaniacal, but otherwise uninteresting, nature of scientific lives then and now. The reputation is all the more unfortunate knowing that librarians are themselves common victims of unflattering characterizations.

In a highly positive response, a number of scientifically more sympathetic librarians have contributed biobibliographical essays for this special issue. The styles of each author are somewhat different. Formats and types of literature allowed in each essay are variable. Reasons for inclusion of one scientist over another and the total number of scientists treated within any piece was entirely at the discretion of each contributor. Each of these authors has a serious contribution to make, and this editor thanks them on behalf of the readers for their efforts.

By way of contrast—and because I am the editor and get to indulge myself a little—my own introduction allows the reader to consider some of the following comical, tragiocomical, or ironic bits of the human trivia of science. They are taken from a wide variety of sources, most of which are noted in the author's more recent books.

Dr. Jekyll and Mr. Hyde. Humphrey Davy (English chemist, 1778-1829). Believed in imbibing chemicals he created in his lab. Died slowly of the cumulative effects of self-poisoning, but discovered laughing gas—nitrous oxide—along the way.

Wash your hands before eating! Alfred Hofmann (Swiss pharmacologist of the 1930's to 1960's). While working in the lab in 1943, Hofmann got his hands wet from chemicals he was processing and felt faint from their fumes. He dutifully noted, however, that his appetite remained excellent. He experienced some rather dramatic dreams while trying to "sleep it off" on the couch at home. With further research, Hofmann found he could avoid the ill effects of synthesizing what turned out to be LSD by wearing rubber gloves and opening a window. (To say nothing about avoiding dishpan hands.)

Found he couldn't work when on a diet. Wolfgang Pauli (Austrian physicist, 1900-1958). Writer's block always disappeared when he cheated. (I have this in common with Pauli.) Won Nobel Prize in 1945. (No such luck for me!)

Eat every bit of it, it's good for you! Vincent Dethier (American insect physiologist, 1915-). When ordering liver for an insect colony in his lab, he specified a "carboy," an amount known to chemical supplies as about twenty gallons. A purchasing director mistakenly delivered a train boxcar load. (This happened at Duke, and reportedly the liver was repeatedly sent to the student dining hall where it was served until the supply was exhausted. Is that when the place became "the home of the Blue Devils"?)

Recommended a change in diet. Carelton Gajdusek (American infectious disease specialist and medical anthropologist, 1923-). Solved the baffling mystery of a brain disease in the South Pacific by linking the habit of eating the brains of one's deceased loved ones with getting the ailment oneself many years later. (Brains are too high in cholesterol, anyway.)

Why not take all of me? William Shockley (American physicist and electronics pioneer, 1910-1990). While justly famed for his technical work, Shockley became notorious for his extreme views on race, intelligence, and eugenics. Before dying he reportedly donated personal specimens to a sperm bank for the production of genius babies. (This author is not sure if his deposit is gaining any interest.)

Cracked safes for a hobby. Richard Feynmann (American physicist, 1918-1988). Became annoyed at all the locked files and vaults involved in the atom bomb project and mastered both trade secrets and psychological aspects of lock design and combination designation. Won Nobel Prize in Physics in 1965. (He knew it in advance, presumably having broken into the safe of the accounting firm of Olafson, Gustavson and Swenson.)

Went parachuting every Sunday morning. Charles Sherrington (English neurophysiologist, 1857-1952). Nobel Prize in Medicine, 1932. (With faith like that, who needs church, right?)

Why you shouldn't diagnose family members or prescribe medicine for yourself. Elie Metchnikoff (Russian-born discoverer of white blood cell's role in fighting infectious disease). Concurred with future wife's explanation that she had an annoying but not particularly serious case of bronchitis, despite the fact that she had to be carried into the church on their wedding day. She proceeded to die of TB whereupon Metchnikoff swallowed so much poison in a dramatic suicide attempt that he vomited it up. Went on to win Nobel prize for Medicine in 1908. Lived to be 71 and endorsed yogurt as the secret to longevity. (If you eat all that yogurt it only seems that your life is a lot longer.)

Behind every great man, stands a supportive wife. Claude Bernard (French physiologist, 1813-1879). His wife led protest marches against his use of animals in experimental medical research. Unsurprisingly, husband and wife later separated. (Who got the family pets?)

Animal lovers, take note! How did John Audubon (American naturalist, 1785-1851) manage to paint such highly accurate pictures of birds and small animals? Hours of patient observation in the field? No, he was a crack shot and killed each and every one of his subjects so as to examine them at leisure. (Makes you wonder about all those famed human anatomy illustrators, doesn't it?)

Major league ingrate. Robert Koch (German plague fighter, 1843-1910). Won Nobel Prize in Medicine in 1905. Having started out as a struggling country doctor, and having his wife raise literally hundreds of rats, chickens, rabbits, and sheep in their cottage for his studies on infectious diseases, he proceeded to dump her for an actress thirty years his junior when he finally became famous. (On the other hand, Koch did become notoriously cranky in his old age. Probably missed the animals.)

Minor league dropout. Joseph Murray (American Surgeon, 1919-). Reluctantly bypassing professional baseball tryouts for the Red Sox, he goes to medical school after Holy Cross. Performs first successful human kidney transplantation. Wins 1990 Nobel Prize in Medicine. (Is this why the Red Sox still need pitching!) . . .

Sometimes, you can do it all. Rosalind Yalow (American medical researcher, 1921-). Raises two kids in a kosher household, wins Nobel prize in Medicine in 1977 for developing radioimmunoassay. (Every time I recount this, my colleagues say, "Sure, daycare was a lot easier to find then!")

Worked in a journal bindery. Michael Faraday (English chemist,

1791-1867). He took home the loose issues to read. Is that why shipments are delayed nowadays? (He succeeded Humphrey Davy, mentioned above, but as far as we know did not eat or sniff the bindery glue.)

If life gives you a lemon. . . . Jean Henri Fabre (French entomologist, 1823-1915). . . . you sell it to pay your school expenses. Fabre financed part of his education as a street vendor of fruit. Became the world's first serious authority on insect behavior. (Presumably on hot days when the inventory had little visitors.)

Payment in advance, please! James Joseph Sylvester (British-American mathematician, 1814-1897). Not betting on the survivability of a new American school that had offered him a position, Sylvester insisted that Johns Hopkins pay his first year's salary in gold before his departure from England. (Clearly, he suffered from the delusion that he was a football coach!)

And you thought it only happened to rock stars and famous librarians! Wilhem Roentgen (German physicist, 1845-1923). He won the first Nobel Prize for his discovery of a workable X-ray set up. While its power for medical diagnosis was readily appreciated, in the early years before the apparatus became widespread, some of the public mistakenly thought Roentgen carried it around with him, or even that he personally had X-ray vision. Despite the fact that he repeatedly disavowed any clinical expertise, the mild-mannered Roentgen was constantly importuned in public for medical diagnoses and wore disguises that his wife made up for him. (Underneath them, as we all know, he wore his Superman outfit.)

Wear the uniform right, or you can't stay in the Boy Scouts! Enrico Fermi (Italian Physicist, 1901-1954). Fermi was one of the world's leading nuclear scientists of the WWII era and a national hero in his native Italy. Despite the fact that he was publicly apolitical and had previously received substantial financial support and honors from the Fascist government, he thought that it might be wise for the safety of his Jewish wife to quietly defect with his family to America. Going to Stockholm to accept the Nobel Prize seemed a perfect opportunity. He never returned. Curiously, Mussolini totally misread the defection. The Duce had sent Fermi a Mussolini-look-alike uniform to wear to the awards ceremony. When Fermi wore the customary evening formal wear instead, the dictator was enraged and banished him. Thinking that Fermi was merely taking his punishment seriously, the Duce was surprised when the physicist did not return when an "all is forgiven" message was later conveyed to the physicist. (Maybe he should have sent flowers.)

Some research is pure crap, while another discovery is ... The ability to synthesize carbon-containing compounds was one of the greatest triumphs of 19th century chemistry. It was long thought that many of these compounds, so abundant in nature, could only be created by living organisms, and never be duplicated in the laboratory. Now this has been done for literally millions of substances in the past century. The communication from Friedrich Wohler, (German chemist, 1800-1882) reporting the first compound synthetically duplicated began: "I can no longer, as it were, hold back my chemical urine. I can make urea without a kidney, whether of man or of a dog!"

If at first, you don't succeed ... Pharmacologists have long been attempting to modify narcotics to get the medicinal benefits without the addictive drawbacks. Nineteenth century German chemist Heinrich Dreser announced that he had gotten around the problems of morphine with a new drug known as heroin. Sigmund Freud seriously thought the answer was cocaine. A British chemist announced a new morphine alternative in 1966, now known as Bentley's compound, after K.W. Bentley, its discoverer. Unfortunately, it turned out to be superstrong. Its only current use is as a sedative for large wild animals. The recommended dose for an elephant is one milligram. (Irate faculty customers may require more.)

For reprints, write to ... Nikolaas Tinbergen (Dutch wildlife behavior expert, 1907-1990). Tinbergen served in the Dutch Army at the outbreak of World War II and was captured by the German Army. He nonetheless managed to publish one of his papers in the prestigious German journal, *Zeitschrift fuer Tierpsychologie (Journal of Animal Psychology)* while in a prisoner of war camp. He remained remarkably free of bitterness and happily accepted the Nobel Prize with two colleague from former Axis partners, Germany and Austria, in 1973. (And your people complain they can't work when the air conditioner isn't functioning.)

Improbable name awards (a tie). Leading British neurosurgeon of the first half of this century: Sir Walter Brain (1895-1966). Leading British expert on insect development, Sir Vincent Wigglesworth (1899-1990).

Forget the nobel prize, the man has figured out a way of handling unwanted mail! Francis Crick (British-American Molecular Biologist, 1916-). The quietest partner of the 1962 Nobel Prize team has developed an all-purpose checklist to answer the flood of mail that celebrity scientists often get. It reads: "Dr. Crick thanks you for your letter but regrets that he is unable to accept your kind invitation to: Send an autograph. Provide a photograph. Cure your disease. Be interviewed. Talk on the radio. Appear on TV. Speak after dinner. Give a testimonial. Help in your project. Read your manuscript. Deliver a lecture. Attend a conference. Act as chairman. Become an editor. Write a book. Accept an honorary degree. (What, no contingency for mail from the publisher's sweepstakes with Ed McMahon's picture on it?)

Why it seems that there aren't enough science librarians with good scientific backgrounds. In 1974, this guest editor was in the middle of his last semester of library school and intent on becoming a cataloger in the humanities. However he then fell down in front of a bar, breaking his leg rather exquisitely. He missed several weeks of school owing to a hospital stay and remained unable to drive to library school because of a cast that extended from ankle to waist. Motivated by a need to make up for the lapsed fellowship income from library school, he nonetheless managed to ascend the thirty stone steps to Holy Cross College's library on crutches to interview for a temporary position as Science Librarian that had unexpectedly cropped up at that institution. The head librarian, known to be an exceptionally wise and even saintly man, sympathetically explained that this editor would not be suitable for the job, since the Science Library had three balconies of stacks overhanging the reading room, and it was clear that the necessarily frequent accompanying of customers to those stacks would be impossible with such a cast. Frantic, this editor nagged the surgeon into cutting off the cast way ahead of schedule. When the now unencumbered young librarian returned for another interview only a day or two later, the head librarian noted that someone who had rashly circumvented medical advice might not be the wisest person to handle important scientific information. Momentarily flustered, but then suddenly hitting upon a response, your resourceful editor blurted out: "Sir, it was due to a pilgrimage to Lourdes!" (Most head librarians expect miracles of their science librarians only after they've hired them.)

Tony Stankus
Editor

Mathematical Biographies:
Profiles and Sources of Information
on Eighteen Mathematicians

Virgil Diodato
Michael Tolan

INTRODUCTION

Here are eighteen people who have done influential mathematics. Some of the eighteen we have known for many years, probably first encountering them in our undergraduate years or even earlier than that. Others are very new acquaintances, people whose lives and works we knew little about until we came across them during the research for this collection.

Selection of the eighteen mathematicians is a personal one. There is no attempt to balance nationalities or genders or mathematical fields. These eighteen seemed more intriguing than many of the other mathematicians who could have been included. The content of the profiles focuses on such qualities as teaching skills; writing style; pioneering spirit; mathematical creativity; the ability to remember what it meant to be a student; and the awareness of the world outside of mathematics. The great mathematician probably has many of these qualities. More than anything, each of these eighteen mathematicians is very human, and we hope the reader will

Virgil Diodato (Assistant Professor) and Michael Tolan (Graduate Assistant/ Student) are at the University of Wisconsin-Milwaukee School of Library and Information Science, P.O. Box 413, Milwaukee WI 53201. Diodato has a PhD in Library and Information Science and an MS in Mathematics. Tolan has an MA in Social Philosophy and Public Policy and is completing a Master's degree in Library and Information Science.

find that humanness intriguing enough to want to read more or want to invite others to read about some of these people.

We have included individuals about whom there is recent (1980-1990) biographical material. Each profile consists of information about the mathematician himself or herself and a brief mention of the biographical source material.

MUHAMMAD IBN MUSA AL-KHWARIZMI –
Flourished 825:
The Namesake

This Islamic mathematician hastened the establishment of the very number system we use today, and his name has been transformed by a millennia of etymological evolution into a common mathematical term.

Al-Khwarizmi produced the first of the great Islamic arithmetics. His *Book of Addition and Subtraction According to the Hindu Calculation* told his contemporaries about the Hindu invention of positional notation and the zero. Arabs brought what we now call the Hindu-Arabic notation with them to Spain in the early eighth century. Later, the translation of Al-Khwarizmi's book into Latin in the twelfth century probably had more of an effect on the spread of the notation through Europe than any other work of that time.[1,2]

We speak Al-Khwarizmi's name today whenever we say "algorithm." Eves tells us that an 1857 Latin translation of his arithmetic corrupted his name to Algoritmi, "from which, in turn, was derived our present word *algorithm*. . . ."[3] We also commemorate Al-Khwarizmi when we speak "algebra," for that word comes from "al-jabr" in the title of his book on algebra, *Kitab al-jabr wa l-muqabala (The Book of Restoring and Balancing)*.[4]

Al-Khwarizmi worked in astronomy and cartography. Being a good, practical scientist, he improved on the works of others. Many of us have heard how Eratosthenes of Alexandria cleverly estimated the circumference of the earth in the second century B.C. (By the way, Eratosthenes was the first Librarian of the Alexandrian Library.) Islamic scientists knew of this measurement, but in their translations of Greek works it was unclear how to transform Greek standard measures into their own. So, Al-Khwarizmi and his colleagues implemented a suggestion by the Caliph al-Ma'mun and reproduced Eratosthenes' survey on a large level area known as the Sinjar Plain. (Their results came to 56 2/3 miles per degree of meridian.)[5]

Al-Khwarizmi's best known geographic work was *The Image of the Earth*, a world map that improved on the maps then in use, by, for example, providing a more realistic shape of the Mediterranean Sea than Ptolemy had offered.[6,7]

Al-Khwarizmi's contributions in arithmetic, algebra, astronomy, and geography were not very original, for his publications borrowed from or were revisions of earlier Islamic, Hindu, and Greek material. Yet his work was "uncommonly influential,"[8] for he introduced this information to the Islamic world, which eventually helped transmit it to a Europe that was awakening from the Dark Ages.

Berggren's *Episodes in the Mathematics of Medieval Islam* devotes part of its first chapter to biographical material on Al-Khwarizmi and several other Islamic scientists.[9] The article on Al-Khwarizmi in the *Dictionary of Scientific Biography (DSB)*[10] provides more detail than Berggren. Both works cite Al-Khwarizmi's great influence, but the *DSB* is more doubtful of the intrinsic value of his work than is Berggren.

RICHARD BELLMAN —
1920-1984:
The Writer

Mathematicians have a wonderful sense of humor. If you had asked American Richard Bellman about himself, he might have told you this story:

When one writes a book, one receives a form from the publisher which will be used for promotion purposes. It always contains some space of noteworthy facts. With tongue in cheek, I always put down the following two pieces of information: (1) I had won the table tennis tournament of Brooklyn

College in 1941. (2) I was the president of the first Jane Russell fan club of Bellville, Swansea, and East St. Louis in 1942. As far as I knew, this important information was never used.[11]

So wrote one who "will be remembered as a giant of the mathematical sciences of this century. Throughout the world, he was considered America's most important mathematician. He personally knew and was respected by mathematicians in almost every country."[12] Bellman worked in many areas, including computer science; management science and operations research; fuzzy sets; and control theory. This resulted in some 650 research papers and 40 books.

In his autobiography, *Eye of the Hurricane*, Bellman lets us into the maturing mind of a mathematician. For example, his "first piece of good mathematics" was done at the University of Wisconsin—Madison in the early 1940s. While working on his master's degree, Bellman had a "simple" procedure for doing mathematics. He would go to the Mathematics Library at Madison and scan recent issues of journals and look for something interesting. One such paper in the *Duke Mathematical Journal* or the *Bulletin of the American Mathematical Society* (he did not remember which journal it was) on stability raised a problem that Bellman within an hour turned into an inequality. His result discouraged him, because his "previous experience with inequalities had been that they produce nothing useful." Yet his persistence and openmindedness made him check and recheck his result, and today it stands as the Bellman-Gronwall Inequality.[13]

This work also eventually led to his dissertation, which was published in 1953 as the now "classic in the field" *Stability Theory of Differential Equations*.[14] This was the first of his many books: "I enjoy writing, but recognize it involves work. . . . Writing is like playing the piano or tennis, it involves constant practice."[15]

As Professor of Mathematics, Electrical Engineering, and Medicine at the University of Southern California, Bellman saw applications of computers and mathematics to the flow of drugs in the human body. He himself "labored under severe disability" in later life and devoted much of his time then "to a project designed to bring microcomputers into the institutions and homes of the handi-capped. . . .''[16]

In addition to his interests in health care delivery, he was deeply concerned and wrote papers on the environment and the societal and human aspects of the computer.[17] Thus he saw (one biographer says he ''was almost obsessed with'')[18] how applied mathematics can solve problems of mankind and improve our quality of life.

This prolific mathematician received many awards and honorary degrees. In memory of him, entire issues of journals were devoted to his work, including the June, 1986 issue of *Computers & Mathematics with Applications* and the October, 1986 issue of the *Journal of Mathematical Analysis and Applications*.

Why read a mathematician's autobiography? As in Bellman's *Eye of the Hurricane*, an autobiography often shows us that there is a very identifiable experience that the individual remembers as the first time he or she made an original contribution to mathematics. Pick up one of the autobiographies mentioned in these profiles, find that first wonderful experience, and you have encountered the high point of the book.

DAVID BLACKWELL —
1919- :
The Teacher and the Student

When you think about Blackwell, you think about teaching and learning. This American mathematician has a fine reputation as a teacher. In one interview, he made so many trips to the blackboard that the interviewer exclaimed, ''You must like to teach.''[19] Even as a senior professor, he still has taught both very advanced and very basic courses.

Blackwell's students have the advantage of working with a man who is keenly aware of what it means to be a student. For example, do students sometimes doubt themselves? Blackwell remembers that he once doubted himself. He ran into the nagging question that many new doctoral students encounter: Can I do it?

> [A]fter I got my master's degree I suspect that if I had got-
> ten a job as a high-school mathematics teacher I would have
> taken it. . . . You go to college for four years and you go out
> and you get a job. Some people go on. I knew I could do the
> course work . . . but I didn't know whether I could write a
> thesis. Does anyone really know whether he can write a thesis
> until he does?[20]

Do students earn degrees, take on jobs, and only then decide
what they want to do with the degree? Although Blackwell is famous for his work in statistics and related areas, he says that his
interest in statistics grew out of a talk he heard not before but after
he obtained his PhD![21]

Do students (and even professors) get involved in social causes?
Being a professor at the University of California-Berkeley for many
years means he was there during the student activism of the 1960s.
Blackwell was active. For example, he signed a letter written by
statistical colleague Jerzy Neyman soliciting funds for Dr. Martin
Luther King's Southern Christian Leadership Conference, " 'to help
the young generation of Negroes in their struggle to attain the status
of citizens in the United States of America.' "[22] He participated in
other fund raising activities for such groups, and he joined in a
faculty protest of the suspension of eight Berkeley students during
the Free Speech Movement there.[23]

Are some students leaders? Blackwell was president of the mathematics club while a student at the University of Illinois.[24] It is not
surprising that later as a mathematician he took on similar responsibilities. He was president of the Institute of Mathematical Statistics
and second Chairman of the Department of Statistics at Berkeley.[25]

Besides all this, Blackwell was a leading statistician at one of the
world's centers of statistical research. As Paul Halmos put it,
Blackwell has "become perhaps the best black mathematician
alive."[26] His success came despite the impediments created by
race-conscious society when he was a young student and teacher.
Blackwell's race was the only reason he was not hired onto the
Berkeley campus in the 1940s rather than the 1950s.[27]

There is an interview with Blackwell in Albers and Alexanderson's wonderful collection, *Mathematical People: Profiles and In-*

terviews.[28] Blackwell appears frequently in Reid's biography of colleague Jerzy Neyman, *Neyman: From Life*.[29] (If you read enough mathematical biographies, the lives intersect!) The reader also may look at Newell and others' *Black Mathematicians and Their Works*. This collection contains 26 research papers by Black mathematicians, including one by Blackwell. There are *Who's Who*-like entries for 62 individuals. The Preface briefly discusses several pioneering Black mathematicians.[30]

GEORGE BOOLE – 1815-1864: The Teacher of Logic

We remember Irish mathematician George Boole for his algebra, his logic, and perhaps even his polynomial. However, what is too easily forgotten is that he must have been an excellent teacher, one who was deeply committed to his students. The story of his death epitomizes this. Whether or not apocryphal, one day in 1864 Professor Boole walked in the rain and lectured while wet, rather than miss class at Queens College, Cork. Since he already was in bad health, this soon led to his death.

Boole saw from an early age that teaching was to be his lifetime endeavor, though his original motivation may have been more economical than inspirational. At the age of 15 or 16 he selected teaching rather than holy orders and taught mathematics to provide money for his family. By 20, he had established his own school. At 34, his already many years of teaching experience helped him obtain the professorship of mathematics at Queens College, even though he did not have any university degrees.

Boole's algebra created a link between Aristotle's and Gottfried Leibniz's dreams of discovering the rules of human thought and Gottleib Frege's and Bertrand Russell's insights about the very foundations of mathematical thinking. Boole's contribution was to mathematicize or axiomatize logic. That is, he used symbols to represent logical operations, such as ANDing and ORing, and he developed procedures, looking very much like algebraic procedures, that allowed for the purely abstract study of mathematical logic.

It was appropriate that his book about all this was boldly called

An Investigation of the Laws of Thought (1854). Perhaps Boole today would have been a teacher of artificial intelligence and cognitive science. In the first paragraph of Chapter I of *Laws of Thought*, we read:

> The design of the following treatise is . . . finally to collect from the various elements of truth brought to view in the course of these inquiries some probable intimations concerning the nature and constitution of the human mind.[31]

The respect of others led the young teacher to be selected at the age of 19 to give the address marking the presentation of the bust of Sir Isaac Newton at the Mechanics' Institute. He received the Royal Society of London's Royal Medal in 1844 and was elected a fellow of the Royal Society in 1857.

Boole is so well known that this profile draws from several general sources, *Asimov's Biographical Encyclopedia of Science and Technology*,[32] *A Biographical Encyclopedia of Scientists*,[33] and *Dictionary of Scientific Biography*.[34] Boole is found in many collective biographies of mathematicians, including those by Hollingdale (*Makers of Mathematics*)[35] and Ashurst (*Founders of Modern Mathematics*).[36] Individual works on Boole include Desmond's *George Boole, His Life and Work*.[37]

GEORG CANTOR –
1845-1918:
The Revolutionary

We have heard that this German mathematician helped revolutionize modern mathematics, but was he also the Vincent Van Gogh of mathematicians? Dunham makes this comparison between Van Gogh and Georg Cantor:

> The two men had a certain physical resemblance. . . . Both were much drawn to artistic enterprises, enjoyed literature, and wrote poetry. . . . Both men were extremely intense, exhibiting a tremendous devotion to their chosen work. And, of course, both men suffered from mental problems that not only

saw them institutionalized but also weighed upon their minds as they contemplated recurring attacks in the future. . . .

Most of all, both Van Gogh and Cantor were revolutionaries. Just as Vincent in his brief and turbulent career managed to carry art beyond its impressionist boundaries, so too did Cantor move mathematics in profoundly new directions.[38]

The new directions included the founding of set theory. Some authors even ascribe a specific date to its founding: December 7, 1873! On that day Cantor wrote to Richard Dedekind that he had proved the real numbers to be uncountable.[39] This was a fascinating, bold, and profound result. Previously, mathematics had not probed the infinite very deeply. Cantor now showed that there were at least two kinds of infinities. The positive whole numbers (1, 2, 3, . . .) provided an example of a well-known, or in Cantor's word, "countable" infinity. Cantor's new, uncountable infinity described the real numbers, which include irrational numbers like the square root of 2. The reals, in a sense, are more infinite than the whole numbers are.

That is not all. Cantor produced something even more radical to the mathematics of his day. He found not merely two kinds of infinities; he claimed there to be many kinds. There is an infinity of infinities! Perhaps his deep religious feelings made that finding less difficult for Cantor to accept than for many of his contemporaries.

There were many denunciations, especially by Leopold Kronecker, whom Eves describes as Cantor's "harsh and relentless critic."[40] Today Cantor's work is seen as fundamental in the most profound sense, and pervades so many areas of mathematics.

Georg Cantor takes up most of the final two chapters in Dunham's *Journey Through Genius: The Great Theorems of Mathematics*.[41] The book is not a collection of biographies, but Dunham's graceful description of the work of some of the greatest mathematicians gives us insight into their lives. There is half a chapter on Cantor in Hollingdale's *Makers of Mathematics*[42] and a full chapter in Ashurst's *Founders of Modern Mathematics*.[43] A respected individual biography is Dauben's *Georg Cantor: His Mathematics and Philosophy of the Infinite*.[44]

CHENG DAWEI —
Flourished 1530:
The Honored Elder

We sometimes hear that a great mathematician often does his or her most creative work at an early age. Historians and psychologists of mathematics can debate the truth of that, but there are fine mathematicians who have produced their best work in old age. The Chinese mathematician Cheng Dawei wrote his greatest work at the age of 60. His *Systematic Treatise on Arithmetic* is a collection of some 600 computational problems. As Euclid is important for codifying and adding to the works of his ancient predecessors, Cheng Dawei's contribution was to gather problems set out in China by the mathematicians who lived in the fifteenth century. His book teaches the reader how to perform calculations, which are usually very practical, such as those needed in the commerce of cereals and cloth, in construction, and in the collection of taxes.

For some 150 years, Dawei's book was the most influential work in Chinese mathematics. It held a place equal to that of the Four Books and Five Classics, which "form the canon of classical Chinese literature and occupy a place in Chinese culture akin to the Old Testament in Western cultures."[45]

Cheng Dawei's book also contains autobiographical material. Dawei tells his Chinese readers that his arithmetic book grew out of his lifetime dedication to mathematics. He found as much refreshment, peace, and hard work in it as he did in the rivers that he had plied as a young trader.

> I started to learn mathematics when I was young. About the age of 20 I was trading and travelling around the Wu and Chu districts and sought out expert teachers. Then I retired and studied very hard near the source of the river Shuai for more than 20 years in order to collect together the fundamentals until at last I felt I had achieved something. So I put together the methods of the various schools and added my own ideas and having got that all together I wrote a book.[46]

Crossley and Lun's translation of Yan and Shiran's *Chinese Mathematics: A Concise History* contains only a few biographical entries. Yet it devotes almost four pages to Cheng Dawei.[47] One of the strengths of the book is the collection of excerpts from original sources.

PAUL ERDOS—
1913- :
The Traveller

Paul Erdos is "The Peripatetic Mathematician,"[48] the "world's most prolific mathematician [who] has no home, no checkbook, no job—only an infinite obsession."[49] Erdos lives to travel and to share his ideas; very possibly he is the most travelled and most prolific mathematician ever. He flies from country to country, giving lectures, participating in symposia, visiting mathematics departments and walking from office to office. Alexanderson's interview caught Erdos along the way from Canada and Mexico to Texas, Florida, Tennessee, Zurich, Budapest, and India, while Tierney talked with him as he stopped between North America and a trip to Israel, Germany, Hungary, and Australia.

He travels in space so that he can travel in mind. He can jump amazingly from problem to problem. As of 1984 his 1,000-plus publications included papers with at least 250 different coauthors.[50] His visits to individual mathematicians all over the world (sometimes announcing "My brain is open")[51] allows this fertile mind to share his mathematical gift with many colleagues. Half-jokingly, colleagues speak of a given mathematician's Erdos Number. The Number is one 1 if the mathematician has co-authored a paper with Erdos; it is 2 if the mathematician has written a paper with someone else who has co-authored a paper with Erdos; and so on. For example, Einstein has an Erdos Number of 2 because he and Erdos share at least one common co-author.[52]

Erdos' major interests have been in number theory and combinatorics. His life is mathematics and the sharing of it with his colleagues.

> I never wanted material possessions. There is an old Greek saying that the wise man has nothing he cannot carry in his hands.[53]

Or in his mind.

Because Erdos is always on the move, the reader will have to be satisfied with the brief interviews that he has granted. Tierney's interview is in *Science84*,[54] while Alexanderson's is another gem from *Mathematical People: Profiles and Interviews*.[55]

EVARISTE GALOIS – 1811-1832: The Young Man

He was always young. What we remember is that the French mathematician Galois died so young and yet has had an impact on mathematics for the century a half since. Perhaps older biographies like Infeld's *Whom the Gods Love*[56] have made Galois's life too romantic.

Perhaps Galois's life and death have tricked us into believing that good mathematics must be done only by the young.

Whatever the link, if any, between age and mathematical ability, Galois indeed was a great, young mathematician. As Eves says, "Galois essentially created the study of groups."[57] He used his inventions in group theory to make discoveries in the theory of equations. Very little of his work was published during his short life. Over the next century, there eventually appeared his memoirs and papers as well as enrichments of his ideas by individuals like Augustin-Louis Cauchy, Camille Jordan, Felix Klein, and Sophus Lie.[58]

Is the life and work of Galois inspirational or romantic because he died young? No. We are moved because he overcame so much to communicate his work to us. He had to try several times before being accepted into formal higher education; his father strongly encouraged his mathematical work but the elder Galois committed suicide when Evariste was eighteen; Evariste's political fervor during the French Revolution eventually landed him in prison two years later.[59]

The greatest barrier was Galois getting himself involved in a pistol duel that he knew he could not win. Shortly before he died he wrote several testaments. One was a letter to Auguste Chevalier in which Galois summarized the mathematical work that he had been doing – much of it not yet completely developed. During this preparation for death, he scribbled on one of his unpublished mathematical papers, probably in response to an incomplete proof, "I have no time."[60] The mathematicians of the next hundred and fifty years would take the time to work on these fertile ideas.

The romantic flavor of Galois's life was emphasized in Infeld's *Whom the Gods Love*. It is an historical novel that includes an Afterword that sorts the truth from fiction, for there is "[m]uch truth and some fiction . . . mixed together" in the work.[61] Originally published in 1948, the book was later distributed by the National Council of Teachers of Mathematics as a reprint in their Classics in Mathematics Education Series. Galois appears in many collective works, including Ashurst's *Founders of Modern Mathematics*.[62]

KURT GODEL —
1906-1978:
The Selective Friend

The work of Austrian-American logician and philosopher Kurt Godel was "singular and monumental;"[63] "remarkable" and "fundamental;"[64] "startling;"[65]; and had "enormous repercussions."[66]

Being a very private person and very judicious about what he published, Godel was not well known outside mathematical and philosophical circles during his life. Hofstadter's fascinating (and "surprisingly popular")[67] book, *Godel, Escher, Bach: An Eternal Golden Braid*[68] opened up Godel to the rest of the world after his death. Hofstadter tells us about Godel's Incompleteness Theorem, one of his major achievements. In the words of John von Neumann, the Theorem "demonstrated the existence of *undecidable* mathematical propositions. . . . It will never be possible to acquire with *mathematical means* the certainty that mathematics does not contain contradictions."[69]

Godel showed that it is impossible to prove that a formal, axiomatic, deductive system is consistent, using the methods of that system. This result applies to axiomatic systems regardless of whether they belong, say, to mathematics or philosophy. What an impact that must have had in 1931! After all, much mathematical energy had been spent in the late nineteenth and early twentieth centuries debating the logical foundations of mathematics. It has been suggested that as knowledge and analysis of Godel's work has spread, the impact on mathematics, philosophy, and other fields is increasing and may continue to increase for hundreds of years.[70] The Incompleteness Theorem is only one of several major findings of Godel.

Godel's success followed from his lifelong dedication to fundamental theoretical work. His commitment was so strong and so wise that he selectively removed outside distractions. He had only a few friends, but they were "beautiful" friendships; he published little, but "concisely" and with great impact; he partook in only a "few lectures," a "few invitations," and little travel. But "when he does make a commitment (to others), he is conscientious about fulfilling it to perfection. . . ."[71]

We should not misunderstand this selective exclusion of distractions. If he was a recluse, he still had deep feelings about people. One biographer sees the friendship between Godel and Albert Einstein as one of the greatest in the history of science. If he was a pensive philosopher, he cared strongly about the world's "strifes, struggles, conflicts, differences, disagreements, contradictions, and inequalities."[72]

How unique was Godel? Well, how many mathematicians have classical music or poetry devoted to them? Henze's 2nd Violin Concerto was inspired by "Hommage a Godel," a poem created by Enzensberger. An English translation of a couple of stanzas:

> You can describe your own language
> in terms of your own language:
> almost, but not quite.

> You can fathom your own brain
> by means of your very own brain:
> almost, but not quite.
> etcetera.[73]

Wang's *Reflections on Kurt Godel*[74] draws from Godel's unpublished as well as published work. It devotes as much space to philosophy as to mathematics, and that is appropriate. Part I, Facts, may be more accessible to most of us than the more technical Part II, Thoughts, and Part III, Texts.

PAUL HALMOS—
1916- :
The Master of Words

If you select just one of the mathematicians mentioned in this article to read more about, we recommend American Paul Halmos. In our search for mathematicians' biographies, the "automathography" by Halmos seems to have gotten checked out more times per year of its short existence than any other item we have looked at. It is no wonder, for this work is so enticing. As Halmos says of himself, "I like words more than numbers, and I always did."[75] In his wonderful words he helps us realize that although mathematicians have skills that may be very difficult to understand, they are understandably very human, too.

Halmos has done his best known research in algebraic logic, Hilbert spaces, and measure theory. But when he entered college, the University of Illinois, at the age of 15, he registered as a major in chemical engineering. As is true for many college students, he "shilly-shallied" about his major. By the end of his first year, he realized chemistry was not inspiring him at all, and he switched to a general liberal arts major with the intent of selecting either philosophy or mathematics. As it turned out, the degree (completed in three years) was in mathematics, though he also satisfied the requirements for a degree in philosophy. One of the undergraduate teachers who impressed him most was Harry Levy. "Above and beyond the call of duty he taught me library technique," taking him to the mathematics library and pointing out the important journals and other sources of information. Halmos now does that for some of his own students.[76]

Halmos went on to receive his PhD at Illinois (after having flunked the oral comprehensive exam in an abortive attempt to get a master's in philosophy). His first mathematical research at Illinois was inspired by R. D. Carmichael. It answered the question: which expressions of the type $ax^2 + by^2 + cz^2 + dt^2$, where a, b, c, d are positive integers, can represent every positive integer with one exception?

I found 88 candidates, proved that there could be no others, and proved that 86 of them actually worked. (Example: $x^2 + y^2 + 2z^2 + 29t^2$ fails to represent 14 only.). . . . No inspiration was needed for this, my first published research, only patience and diligent application of the techniques Carmichael taught me, but the work gave me a feeling of accomplishment and (badly needed) the confidence that I could do research. I was very proud. I bought 200 reprints; it took me many years to find enough people to give them to.[77]

The excitement of that experience happened again and again in Halmos' mathematical life. You can read about his bouts with analysis (he "finally mastered epsilons"); his exhilarating introduction to set theory (he "was like a child with a new toy: I thought it was beautiful and astonishing");[78] his inspirations from other mathematicians, especially his PhD advisor, Joe Doob ("My eyes were opened. I was inspired. He showed me a kind of mathematics . . . that wasn't visible to me before"), and John von Neumann ("his speed, plus depth, plus insight, plus inspiration turned me on").[79]

There is so much to learn from Halmos. Read his automathography and learn how to think, how to supervise, how to advise, how to be a chairman, how to write mathematics, how to teach, how to study.

Here you sit, an undergraduate with a calculus book open before you, or a pre-thesis graduate student with one of those books whose first ten pages, at least, you would like to master, or a research mathematician (established or would-be) with an article fresh off the press — what do you do now? [S]tudy actively. Don't just read it; fight it! Ask your own questions, look for your own examples, discover your own proofs. . . . I keep active . . . by changing the notation. . . . I get a feeling of being creative, tiny but non-zero even before I understand what's going on, and long before I can generalize it, improve it or apply it, I am already active, I am doing something.[80]

Halmos can speak eloquently on mathematics and make the essence of it accessible to the nonmathematician. In *Mathematical*

People: Profiles and Interviews, Albers asked him what mathematics meant.

> [Mathematics] is security. Certainty. Truth. Beauty. Insight. Structure. Architecture. I see mathematics, the part of human knowledge that I call mathematics, as one thing—one great, glorious thing. Whether it is differential topology, or functional analysis, or homological algebra, it is all one thing. . . . They are intimately interconnected, and they are all facets of the same thing. That interconnection, that architecture, is secure truth and is beauty. That's what mathematics is to me.[81]

His automathography is called *I Want to Be a Mathematician: An Automathography*.[82] If you have not time to read it, you can get a good sense of Halmos by looking at one of his articles about mathematics, which have appeared in such publications as *American Scientist*[83] and the *Two-Year College Mathematics Journal*.[84]

FELIX KLEIN—
1849-1925:
The Universalist

Yes, this is the Klein of the "Klein Bottle," that strange structure that has no inside and no outside, with all one surface and no edges. But the emphasis in this profile will be less on his work and more on his influence on mathematics education in Germany.

Klein saw higher education as an opportunity for a student to become familiar with many areas of knowledge. He himself worked in diverse fields. His entry in *James & James Mathematics Dictionary* demonstrates this. Whereas most mathematicians are described in one word (algebraist, geometer, topologist, for example), Klein's entry is unusually long. He was: "algebraist, analyst, geometer, topologist, mathematical historian, and physicist."[85]

His universalist philosophy gave Klein a mission: remind the mathematical world that applied mathematics indeed was a worthy branch of mathematics and should not be forgotten in the rush to solve the wonderful problems of pure mathematics. One way to do this in late nineteenth century Germany was to advocate the study of applied mathematics in universities, alongside pure mathematics, rather than isolate applied mathematics in technical colleges. This would mean integration of students pursuing pure mathematical-philosophical studies with those planning to be engineers and technicians.[86]

His suggestion, in 1888, to integrate technical colleges and universities, was "met with violent opposition from the engineering community"[87] Despite his strong opinions about integration, he had the maturity and political good sense to, by 1900, change his mind. Instead of changes in institutions, he then was proposing curricular change at the high school level. With the blessings of the Ministry of Education, Klein oversaw the introduction of mathematical reforms into German high schools in the early 1900s. These measures improved the coordination of curricula between high schools and technical colleges; they encouraged mathematics majors to obtain secondary school teaching licenses from the technical colleges; and this led to requirements that applicants for teaching licenses be able to apply mathematics to fields such as astronomy, meteorology, physics, and statistics.[88]

Readers in the United States of the 1990s should find that Klein was as concerned as they are about the mathematics curriculum in high schools and about teacher education. Imagine: here is a great mathematician of his day stepping in and doing something about the problem!

Schubring's chapter in Rowe and McCleary's *The History of Modern Mathematics*[89] is not a biography but rather a description of Klein's impact on German mathematics education. The chapter gives information about Klein's life not easily found elsewhere. There is even a list of courses that Klein took during his broad training in the natural sciences and mathematics. There also is the work by Yaglom, *Felix Klein and Sophus Lie*.[90]

SONJA KOVALEVSKAYA —
1850-1891:
The Networker

Kovalevskaya was "the only Russian and the only woman and the only" product of the 1860s age of intellectual nihilism "involved in the late nineteenth-century Western European mathematical world.''[91]

Koblitz tells us that intellectual nihilism in Russia of the 1860s took the expression of great faith in the natural sciences, including mathematics, and a belief ''that men had a moral obligation to aid'' women to ''educate themselves so they could help the masses of ordinary Russians.'' It is not surprising then that Kovalevskaya was all her life a women's rights advocate and political progressive.[92]

In 1870 Kovalevskaya, because she was a woman, was denied entrance into the University of Berlin. Karl Weierstrass taught at the University and took on Kovalevskaya as a private pupil during 1870-1874. In the final year with Weierstrass, Kovalevskaya:

> was awarded, in absentia, the degree of Doctor of Philosophy by Gottingen University [the first mathematics doctorate awarded to a woman in modern Europe] and, because of the outstanding quality of a submitted paper on partial differential equations, was excused from taking the oral examination.[93,94]

Kovalevskaya was a great mathematician and a great communicator. She took on the role of gatekeeper, establishing communication links between the Russians, who preferred the more concrete attack on theoretical problems, and the Europeans, who enjoyed the abstract methods of analysis. Her knowledge of Russian and European mathematics and mathematicians allowed her to become a hub of a network among the French, German, and Russian schools of mathematics of the late nineteenth century. Also of vital importance in this network was her editorship of *Acta Mathematica*, through which her Russian colleagues saw her as a representative to the West.[95]

Her own work dealt with analysis and with mechanics and mathematical physics. She is especially remembered for the Cauchy-Kovalevskaya Theorem in partial differential equations and work on the revolution of a solid body about a fixed point.[96]

Two individual biographies on Kovalevskaya (also known as Sophia and as Kovalevsky) appeared in 1983. We were drawn to Koblitz's *A Convergence of Lives*; the other is Kennedy's *Little Sparrow*.[97] She is in collective biographies, including Alic's *Hypa-*

tia's Heritage[98] and Abir-Am and Outram's *Uneasy Careers and Intimate Lives*, which includes an article by biographer Koblitz.[99]

ADA BYRON LOVELACE –
1815-1852:
The Persistent Partner

British mathematician Lovelace was persistent. Why? Because she had to keep after herself to learn mathematics and to keep after Charles Babbage to do *something* with his fascinating ideas. "Ada was an extremely ambitious women who, above all else, longed to become a famous scientist" and who was much more exuberant and less humble than other women scientists of her day.[100] The big problem in attaining her goal was overcoming barriers to her own education. Seemingly simple matters such as acquiring a tutor and using libraries were not easily dealt with, despite her family's high place in society. Overcame them she did, and it is interesting that one important access to libraries was through her husband, William King, Earl of Lovelace. As a member of the Royal Society of London, he was able to carry home materials from the Society's library, into which Ada was not permitted to enter.[101]

If you knew only a little about Lovelace, perhaps what you do remember is her link with two famous names: Byron and Babbage. She was the daughter of Lord Byron, the poet, and colleague of Charles Babbage, would-be inventor of the computer.

The link among these three people may well have been creativity. The poet passed this on to his mathematician daughter, who in turn worked with Babbage to develop ideas that were about a century ahead of their time.

Lovelace's best mathematical works were the papers she published in response to Babbage's description of his Difference and Analytical Engines. Some of her papers even dealt with how to program these proposed machines – programs that would get the machines to do calculations and to write music.

Babbage neither had the money nor access to twentieth century engineering to implement his ideas, and Lovelace dogged him about his efforts. Lovelace implored the inventor to keep his mind on his plans for the Engines and to avoid sloppiness in documenting this work. This pair, along with the Earl of Lovelace, even tried (and failed badly) to raise money for the Engines by betting on horse races.[102]

Nevertheless, in the mathematics of nineteenth century England it was only she and Babbage who:

> were making important advances, and they were so far ahead of their time that the significance of their work would not be appreciated for another century.[103]

Alic's *Hypatia's Heritage* compresses a lot of information about Lovelace into one section of a chapter.[104] There is also a chapter for Lovelace in Perl's *Math Equals*.[105] Individual biographies of Lovelace include Stein's *Ada: A Life and Legacy*[106] and Baum's *The Calculating Passion of Ada Byron*.[107]

JERZY NEYMAN –
1894-1981:
The Statistician

Is it okay to include a statistician among these other profiles? Is a statistician a mathematician? These seem like silly questions, but they suggest that mathematics students who are well aware of the great people of their field might know little of the lives of statisticians. Even in Eves' *An Introduction to the History of Mathematics*, there are very few names associated with statistics and probability outside of the obligatory homage to the Bernoulli family.[108]

Reading about Russian-Polish-American statistician Jerzy Neyman can increase awareness of statisticians, especially because his life intersected many other greats in that field.

If you have held one of William Feller's probability textbooks in your hand (or even had to do exercises from it); or worked with formulas or procedures attached to names like A. N. Kolmogorov and Karl Pearson; or used statistical tables with little footnotes to works by R. A. Fisher; or wondered what the "Student" in the t-test is named after (it is a person's name), you should read about Neyman and discover that he worked with or influenced or was affected by these and so many others in the field.

Neyman's life also is interesting because he spent some of his energies as an administrator. In about 1947 he took on the directorship of the Statistical Laboratory at the University of California-Berkeley, the Lab being newly separated from the Department of Mathematics. Thanks to Neyman's leadership, the "capital of statistics, which had once been London, had moved to the New World," specifically Berkeley.[109] In 1955 the Lab evolved into the Department of Statistics, with Neyman as Chairman for 1955-1956. (His work did not end in 1956; between then and his death in 1981, Neyman worked at Berkeley, active in a host of projects.)

Descriptions of Neyman's years as an administrator sound very true to academic life. Problems with budgets, personnel, equipment, rooms, grant writing, recruiting students were all part of the day to day of even one of this century's great statisticians!

But what about his statistics? While in Europe during the 1920s and 1930s he worked with people like Egon Pearson and W. S. Gosset. Egon Pearson's father was Karl Pearson, the Father of Mathematical Statistics. Gosset created the t-statistic and signed some of his papers "Student," doing this perhaps because his superiors at a brewery might not have wanted it known that one of their employees was writing scholarly papers.[110]

Neyman and Egon Pearson's collaboration, buoyed by ideas suggested by Gosset, led to papers that popularized the use of careful sampling; hypothesis testing; and confidence intervals. It may seem strange today, but techniques of statistical analysis so common to us had to be invented. Neyman, Pearson, and Gosset were among the creators of these tools.

Reid's *Neyman: From Life*[111] contains descriptions and photographs of many of Neyman's encounters. The photos especially help us see how human these scholars were.

Neyman's biographer, Constance Reid is a nonmathematician but such an accomplished mathematical biographer that a profile of Reid herself appears in Albers and Alexanderson's *Mathematical People: Profiles and Interviews*.[112]

EMMY NOETHER — 1882-1935: The Pioneer

According to Albert Einstein, Noether was one of the "most significant creative mathematical genius[es]" of her time.[113] She was a pioneer, too. This German mathematician earned her PhD summa cum laude in 1907 from the University of Erlangen, which only three years earlier admitted its first women as regular students, rather than as merely auditors.[114]

Noether helped pioneer twentieth century abstract algebra, working especially with such structures as ideals, noncommutative rings, and representations. However, she did not start out as a pure algebraist.

To be a pioneer you must be in the right place at the right time. Her dissertation dealt not with purely abstract algebra but with algebraic invariants, a topic she tolerated. Why did she select that topic? A big reason probably was that it was of interest to her advisor, Paul Gordon, the "king of invariants."

> Emmy Noether herself later referred to her thesis, as well as to several consecutive papers on the theory of invariants, as "crap." In 1932 she declared that as far as she was concerned, her dissertation was forgotten. . . .[115]

Noether eventually got to the right place, the University of Gottingen. The right teacher was there, algebraist Ernst Fischer, who drew on the work of David Hilbert in abstract fields, rings, and modules.

> [I]n 1919, Emmy Noether states in a curriculum vitae that it was Fischer who had awakened her interest in abstract algebra, approached from the point of view of arithmetic, and that this had determined all her later work.[116]

If one is to be an effective pioneer, then there should be others who follow in the pioneer's footsteps. So it was with Noether. As a professor at Gottingen between 1919 and 1933, she advised the dissertations of the next generation of algebraists. Known as the "Noether School" this group included Max Deuring, Hans Fitting, Otto Schilling, Werner Weber, and Ernst Witt.

However, by 1933 Adolf Hitler was removing suspicious professors from the universities. Being Jewish and perhaps having had leftist students in her apartment, Noether was a "dangerous" individual. She received the following notice in April, 1933:

With reference to [section] 3 of the statutes for professional civil servants of April 7, 1933, I herewith withdraw your permission to teach at the University of Gottingen.[117]

Thus it was that Noether spent the last few years of her career at Bryn Mawr College and at the Institute for Advanced Study, Princeton during 1933-1935. Though she died unexpectedly in 1935, she had time to advise one doctoral candidate in the United States. Noether's death occurred during that candidate's final semester. The student did obtain her doctorate, but the loss of "her motherly thesis advisor" apparently hurt a great deal and the student abandoned her mathematical work.[118]

Dick's *Emmy Noether 1882-1935* is the source for the information in this profile. It contains a biography as well as almost 100 pages of obituaries and memorial addresses. There is also Brewer and Smith's *Emmy Noether: A Tribute to Her Life and Work*.[119]

GIUSEPPE PEANO —
1858-1932:
The Citizen

Maybe you know about Peano's Axioms and the influence he had on the work of Alfred North Whitehead and Betrand Russell. However, this Italian mathematician made contributions to society, too. As is true for so many mathematicians and other scholars, Peano was a responsible citizen who was strongly involved in issues outside his abstract work.

For example, he helped improve teacher education programs for secondary mathematics teachers in Italy. He certainly knew about this area, for even a great mathematician in academia has student advisees. Apparently Peano saw firsthand that too many of his students became mediocre teachers.[120]

Because he saw language as a barrier to scholarly and professional communication, Peano was influential in the interlingual movement of the early twentieth century. He published an interlingual dictionary, was president of an interlingual academy, and even created an international language. This language was Latino sine Flexione (Latin without Grammar). It employed words that had common roots in many European languages. "At the International

Congress of Mathematics at Cambridge in 1912 he tried to give his lecture in [Latino sine Flexione], but was prevented from doing so. . . . "[121] Nevertheless, he was a lifetime supporter of Flexione, Esperanto, and other such languages.

Remember logarithmic tables? We have calculators and personal computer software today to handle much of that work, but during the Great War the tables were vital. At the time, log tables produced in Italy were not precise enough, and the country imported tables made elsewhere. Peano saw this as "expensive and unpatriotic." He became a member of a committee that improved the tables produced in Italy. Citizen Peano also supported the striking cotton workers in Turin in 1906, and he performed actuarial services to help solve pension fund problems in Italy in the early 1900s.[122]

Of course, Peano was a fine mathematician. Mathematics students today often first encounter Peano when they hear his Axioms. These five statements were forerunners of the logical work found in Whitehead and Russell's *Principia Mathematica*. After accepting the five Axioms, one can go on to prove the existence of each of the positive whole numbers. Like Euclid's insightful Fifth (Parallel) Postulate, Peano's Fifth Axiom stood out from the other Axioms. It was a powerful theorem-proving tool, the Principle of Mathematical Induction.

Even if you have not used the Peano Axioms nor spoken Latino sine Flexione, you may have used others of Peano's inventions. Peano created the symbols we use today for the union and intersection of sets. They look like the letter U, respectively standing straight up and upside down. He also created the more esoteric (but very common to mathematicians) upside-down-A and backwards-E for "for all" and "there exists."[123]

Peano gets one of the ten chapters in Ashurst's fine little book, *Founders of Modern Mathematics*.[124] A recent individual biography is Kennedy's *Peano: The Life and Works of Giuseppe Peano*.[125]

SRINIVASA RAMANUJAN—
1887-1920:
The Friend of Hardy

The mathematical life of Ramanujan was as extraordinary as his pure mathematics is beautiful. As a clerk in Madras, India, he did mathematics in his spare time because he had no university degree. Later, between 1913 and 1920 he was able to devote almost every moment to mathematics, most of it at Trinity College, Cambridge. At Trinity he had a tremendous influence on the great mathematician, G. H. Hardy, who saw Ramanujan as:

> the most romantic figure in the recent history of mathematics. A man whose career seems full of paradoxes and contradictions. . . . [H]e was in some sense a very great mathematician.[126]

The University of Madras rescued Ramanujan from the clerk's office and gave him a research scholarship in 1913 to work on his mathematical ideas. His colleagues encouraged him to write to the mathematicians at Cambridge. Ramanujan's letters so intrigued Hardy and his colleague, J. E. Littlewood, that they invited him to come to Trinity in 1914. The letters contained some conjectures that Hardy and Littlewood had never seen before. Some they could not prove. Said Hardy: "They must be true, because if they were not true, no one would have had the imagination to invent them."[127]

Ramanujan's ideas sometimes came to him as flashes of insight. Mathematicians George Andrews and Bela Bollobas say today that these insights came much more frequently to Ramanujan than to the typical mathematician. His first proofs consisted more of intuition and example than strict deduction, and he was likely surprised that a few of his mathematical insights turned out to be incorrect. Much of Ramanujan's early information about mathematics was self-taught. Bollobas marvels: "in mathematics almost everybody has a very good teacher. It's very rare to find an outstanding mathematician who was not taught by anybody."[128]

Between 1914 and 1919 at Cambridge, Ramanujan, sometimes in collaboration with Hardy and Littlewood, produced results in number theory and various other areas of mathematics. He and Hardy developed an amazing formula for the number of partitions of an integer (the number of sums one can find that add to the given integer). Andrews finds the formula unbelievable, for it involves terms, such as transcendental numbers, that seem to have nothing to do with the summing of integers. But that was the kind of insight Ramanujan had.

Ramanujan fell ill and returned to India in 1919. He died, perhaps of tuberculosis, the following year, at the age of 32. In 1988 his wife, Janaki Amal, recalled that both before and after his years in Cambridge, his time at home was completely devoted to mathematics.

> He worked on sums. He didn't do anything else. . . . He wouldn't stop work even to eat. We had to make rice boats for him and place them in his hand. Isn't that extraordinary![129]

Much of the work done in his last year went into what has been called the Lost Notebook. It was "Lost" because its whereabouts was unknown until 1976, when Andrews found it in the Trinity College Library while he was looking for something else. "I was thrilled by it." Andrews reports that the vast majority of the material contained results unknown at the time to mathematicians, results produced with such great insight that Andrews cannot understand the thought patterns that created them.[130]

One of the few mathematical programs in Public Television's NOVA series is *The Man Who Loved Numbers*. This fine production describes Ramanujan as one who "left behind some of the most remarkable and beautiful ideas in the history of pure mathematics."[131] Librarians may be interested to know that another biography, *Ramanujan, The Man and the Mathematician*[132] was written by S. R. Ranganathan, the classificationist and developer of the Five Laws of Library Science. Ranganathan was a professor of mathematics before entering the field of library science.

MARY ELLEN RUDIN— 1924- : The Enthusiastic Topologist

"[A] bubbling enthusiasm for mathematics" is how one interview describes this American mathematician's teaching style.[133]

Seeking to find out more about this teaching skill, we looked through one of Rudin's monographs. Her *Lectures on Set Theoretic*

Topology at first glance definitely does not suggest enthusiasm. Especially to a non-mathematician, the booklet's cover looks as dry as the symbols inside look intransigent. However, upon giving the work a good scan, there appear words like "gibberish," "striking," "foolhardy."[134] Are these technical terms that mathematicians have borrowed from the vernacular? No! Rudin actually is writing sections of this text in good, enjoyable English. The work also shows a fine sense of the history of her topic, and enthusiasm.

Yes, she must be an excellent teacher and a devoted mathematician, to tell us that there is "beautiful work in infinite-dimensional spaces"; that metrization is "the heart and soul of general topology"; and that "abstract space topology is remarkably set theoretic."[135]

Being our contemporary, Rudin does not have a great deal of biographical material written about her. However, you will find a good piece by Albers, Alexanderson, and Reid in the *College Mathematics Journal*. You will learn how she was nurtured by mathematician R. L. Moore, how he recruited her into mathematics on her first day as an undergraduate, how she took a course from him every semester. "I'm a mathematician because Moore caught me and demanded that I become a mathematician."[136]

Read about Rudin's style of doing mathematics, the importance of the counterexample in her mathematical thinking, and the role of women in mathematics. Read about her teacher, Moore, and his almost legendary teaching methods, including helping his students be unadulterated and independent. Rudin says: "At the time I wrote my thesis, I had never in my life seen a single mathematics paper! [Moore] told his students—at least so I have heard—'I don't want you going to the library and reading papers'."[137]

The article on Rudin is adapted from a new edition of *Mathematical People*: Albers, Alexanderson, and Reid's *More Mathematical People*.[138]

Finding information about Rudin helps us realize that journals can be good sources of biographical material. The *College Mathematics Journal* is an example. In the past few years it has published not only the article on Rudin but also a biographical piece on geometer Charlotte Angas Scott,[139] who, Eves tells us, was "the first British woman to receive a doctoral (in any field)."[140] A couple of other periodical sources for biographical articles are *Mathematics Magazine* and *Historia Mathematica*.

REFERENCES

Al-Khwarizmi:

1. Eves, Howard. *An Introduction to the History of Mathematics*. 6th ed. Philadelphia: Saunders, 1990: p. 24.
2. Berggren, J. L. *Episodes in the Mathematics of Medieval Islam*. New York: Springer, 1986: p. 7.
3. Eves, p. 236.
4. Berggren, p. 7.
5. Berggren, pp. 8-9.
6. Berggren, p. 9.
7. Gillispie, Charles Coulston, ed. *Dictionary of Scientific Biography* (16 vols) New York: Scribners, 1970: vol. I, p. 362.
8. Gillispie, vol. I, p. 362.
9. Berggren, pp. 6-9.
10. Gillispie, vol. I, pp. 358-364.

Bellman:

11. Bellman, Richard. *Eye of the Hurricane: An Autobiography*. Singapore: World Scientific, 1984: p. 198.
12. Adomian, G. and E. S. Lee. "The Research Contributions of Richard Bellman." *Computers & Mathematics with Applications* 12A (June 1986):633.
13. Bellman, *Eye of the Hurricane*, p. 74.
14. Adomian and Lee, p. 633.
15. Bellman, *Eye of the Hurricane*, pp. 149-150.
16. Adomian and Lee, p. 634.
17. Bellman, R. and R. Kalaba. "Some Mathematical Aspects of Optimal Predation in Ecology and Boviculture." *Proceedings of the National Academy of Sciences* 46 (1960):718-720.
18. Esogbue, A. O. "Mathematics, Computers and Systems: Tripartite Interconnections. *Computers & Mathematics with Applications* 16 (no. 10/11, 1988):774.

Blackwell:

19. Albers, Donald J. and G. L. Alexanderson, eds. *Mathematical People: Profiles and Interviews*. Boston: Birkhauser, 1985: p. 26.
20. Albers and Alexanderson, p. 22.
21. Albers and Alexanderson, p. 24.
22. Reid, Constance. *Neyman: From Life*. New York: Springer, 1982: p. 264.
23. Reid, p. 269.
24. Reid, p. 138.

25. Reid, pp. 242, 249.

26. Halmos, Paul R. *I Want to Be a Mathematician: An Automathography.* New York: Springer, 1985: p. 45.

27. Reid, p. 182.

28. Albers and Alexanderson, pp. 18-32.

29. Reid. (See note 22 above.)

30. Newell, Virginia K. and others, eds. *Black Mathematicians and Their Works*. Ardmore, PA: Dorrance, 1980.

Boole:

31. Boole, George. *An Investigation of the Laws of Thought*. New York: Dover, 1951: p. 1.

32. Asimov, Isaac. *Asimov's Biographical Encyclopedia of Science and Technology*. New York: Doubleday, 1964: pp. 273-274.

33. Daintith, John and Sara Mitchell, eds. *A Biographical Encyclopedia of Scientists* (2 vols). New York: Facts on File, 1981: vol. 1, p. 92.

34. Gillispie, Charles Coulston, ed. *Dictionary of Scientific Biography* (16 vols) New York: Scribners, 1970: vol. II, pp. 293-298.

35. Hollingdale, Stuart. *Makers of Mathematics*. New York: Viking Penguin, 1989: pp. 342-349.

36. Ashurst, F. Gareth. *Founders of Modern Mathematics*. London: Frederick Muller, 1982: pp. 29-44.

37. MacHale, Desmond. *George Boole, His Life and Work*. Dublin: Boole Press, 1983.

Cantor:

38. Dunham, William. *Journey Through Genius: The Great Theorems of Mathematics*. New York: Wiley, 1990: pp. 279-280.

39. Gillispie, Charles Coulston, ed. *Dictionary of Scientific Biography* (16 vols) New York: Scribners, 1970: vol. III, p. 53.

40. Eves, Howard. *An Introduction to the History of Mathematics*. 6th ed. Philadelphia: Saunders, 1990: p. 569.

41. Dunham, William, pp. 251-283.

42. Hollingdale, Stuart. *Makers of Mathematics*. New York: Viking Penguin, 1989: pp. 358-364.

43. Ashurst, F. Gareth. *Founders of Modern Mathematics*. London: Frederick Muller, 1982: pp. 69-82.

44. Dauben, Joseph W. *Georg Cantor: His Mathematics and Philosophy of the Infinite*. Cambridge, MA; Harvard University, 1979.

Dawei:

45. Yan, Li and Du Shiran. *Chinese Mathematics: A Concise History*. Oxford: Clarendon, 1987. Translated by John N. Crossley and Anthony W.-C. Lun: p. 188.

46. Yan and Shiran, p. 186.

47. Yan and Shiran, pp. 185-189.

Erdos:

48. Albers, Donald J. and G. L. Alexanderson, eds. *Mathematical People: Profiles and Interviews*. Boston: Birkhauser, 1985: p. 83.
49. Tierney, John. "Paul Erdos Is in Town. His Brain Is Open." *Science84* (October 1984):41.
50. Tierney, p. 42.
51. Tierney, p. 41.
52. Albers and Alexanderson, pp. 87, 91.
53. Tierney, p. 47.
54. Tierney, pp. 40-47.
55. Albers and Alexanderson, pp. 82-91.

Galois:

56. Infeld, Leopold. *Whom the Gods Love: The Story of Evariste Galois*. Reston, VA: National Council of Teachers of Mathematics, 1978. Reprint of 1948 edition.
57. Eves, Howard. *An Introduction to the History of Mathematics*. 6th ed. Philadelphia: Saunders, 1990: p. 492.
58. Eves, p. 492.
59. Infeld, pp. 48, 49, 80, 213.
60. Infeld, pp. 288-290.
61. Infeld, p. 303.
62. Ashurst, F. Gareth. *Founders of Modern Mathematics*. London: Frederick Muller, 1982: pp. 1-14.

Godel:

63. Bulloff, Jack J., Thomas C. Holyoke, and S. W. Hahn, eds. *Foundations of Mathematics: Symposium Papers Commemorating the Sixtieth Birthday of Kurt Godel*. New York: Springer, 1969, p. ix.
64. Eves, Howard. *An Introduction to the History of Mathematics*. 6th ed. Philadelphia: Saunders, 1990: p. 635.
65. Kline, Morris. *Mathematical Thought from Ancient to Modern Times*. New York: Oxford University, 1972: p. 1206.
66. Hofstadter, Douglas R. *Godel, Escher, Bach: An Eternal Golden Braid*. New York: Basic Books, 1979: p. 15.
67. Wang, Hao. *Reflections on Kurt Godel*. Cambridge, MA: MIT, 1987: p. 3.
68. Hofstadter. (See note 66 above.)
69. Bulloff, Holyoke, and Hahn, p. ix, quoting John von Newmann.
70. Wang, p. 1.
71. Wang, pp. 3-4.
72. Wang, pp. 2, 3, 233, 236-238.
73. Henze, Hans Werner, *2nd violin concerto, for solo violin, tape, voices,*

and 33 instrumentalists using the poem Hommage a Godel, by Hans Magnus Enzensberger. Translated by Desmond Clayton and Hans Magnus Enzensberger. New York, Schott Music Corp, 1973: p. 15.

74. Wang. (See note 67 above.)

Halmos:

75. Halmos, Paul R. *I Want to Be a Mathematician: An Automathography*. New York: Springer, 1985: p. 3.

76. Halmos, *I Want to Be a Mathematician*, p. 33.

77. Halmos, *I Want to Be a Mathematician*, pp. 40, 42.

78. Halmos, *I Want to Be a Mathematician*, pp. 55, 57.

79. Albers, Donald J., and G. L. Alexanderson, eds. *Mathematical People: Profiles and Interviews*. Boston: Birkhauser, 1985: p. 125.

80. Halmos, *I Want to Be a Mathematician*, pp. 69-70.

81. Albers and Alexanderson, p. 127.

82. Halmos, Paul R. *I Want to Be a Mathematician*. (See note 75 above.)

83. Halmos, Paul. "Mathematics As a Creative Art." *American Scientist* 56 (Winter 1968):375-389.

84. Halmos, Paul. "The Thrills of Abstraction." *Two-Year College Mathematics Journal* 13 (September 1982):243-251.

Klein:

85. James, Glenn. *James & James Mathematics Dictionary*. 4th ed. New York: Van Nostrand, 1976: p. 218.

86. Schubring, Gert. "Pure and Applied Mathematics in Divergent Institutional Settings in Germany: The Role and Impact of Felix Klein." In: Rowe, David E. and John McCleary, eds. *The History of Modern Mathematics. Proceedings of the Symposium on the History of Modern Mathematics. Vassar College, Poughkeepsie, NY, June 20-24, 1989*. 2 vols. Boston: Academic Press, 1989: vol. II, pp. 171-182.

87. Schubring, p. 184.

88. Schubring, pp. 185-196.

89. Schubring, pp. 171-220.

90. Yaglom, I. M. *Felix Klein and Sophus Lie: Evolution of the Idea of Symmetry in the Nineteenth Century*. Boston: Brikhauser, 1988. Translated by Sergei Sossinsky.

Kovalevskaya:

91. Koblitz, Ann Hibner. *A Convergence of Lives: Sofia Kovalevskaia: Scientist, Writer, Revolutionary*. Boston: Birkhauser, 1983: p. xvi.

92. Koblitz, *A Convergence of Lives*, pp. xiii, xiv, 270.

93. Eves, Howard. *An Introduction to the History of Mathematics*. 6th ed. Philadelphia: Saunders, 1990: p. 573.

94. Koblitz, *A Convergence of Lives*, p. 270.

95. Koblitz, *A Convergence of Lives*, pp. xvii, 245-247.

96. Koblitz, *A Convergence of Lives*, pp. 239-240.

97. Kennedy, Don H. *Little Sparrow: A Portrait of Sophia Kovalevsky*. Athens, OH: Ohio University, 1983.

98. Alic, Margaret. *Hypatia's Heritage: A History of Women in Science from Antiquity to the Late Nineteenth Century*. London: Women's Press, 1986: pp. 163-173.

99. Koblitz, Ann Hibner. "Career and Home Life in the 1880s: The Choices of Mathematician Sofia Kovalevskaia." In: Abir-Am, Pnina G. and Dorinda Outram, eds., pp. 172-190. *Uneasy Careers and Intimate Lives: Women in Science, 1879-1979*. New Brunswick, NJ: Rutgers University, 1987.

Lovelace:

100. Alic, Margaret. *Hypatia's Heritage: A History of Women in Science from Antiquity to the Late Nineteenth Century*. London: Women's Press, 1986: pp. 158, 160.

101. Alic, p. 161.

102. Alic, pp. 159, 161, 162.

103. Alic, p. 157.

104. Alic, pp. 157-163.

105. Perl, Teri. *Math Equals: Biographies of Women Mathematicians + Related Activities*. Philippines: Addison-Wesley, 1978: pp. 101-125.

106. Stein, Dorothy. *Ada: A Life and Legacy*. Cambridge, MA: MIT, 1985.

107. Baum, Joan. *The Calculating Passion of Ada Byron*. Hamdem, CT: Archon, 1986.

Neyman:

108. Eves, Howard. *An Introduction to the History of Mathematics*. 6th ed. Philadelphia: Saunders, 1990: pp. 424-428.

109. Reid, Constance. *Neyman: From Life*. New York: Springer, 1982: pp. 214, 218.

110. Reid, pp. 2, 54.

111. Reid. (See note 109 above.)

112. Albers, Donald J. and G. L. Alexanderson, eds. *Mathematical People: Profiles and Interviews*. Boston: Birkhauser, 1985: pp. 268-279.

Noether:

113. Dick, Auguste. *Emmy Noether: 1882-1935*. Boston: Birkhauser, 1981. Translated by H. I. Blocher: p. 93.

114. Gillispie, Charles Coulston, ed. *Dictionary of Scientific Biography* (16 vols) New York: Scribners, 1970: vol. X, p. 137.

115. Dick, p. 17.

116. Dick, p. 30.

117. Dick, p. 75.

118. Dick, pp. 85-86.

119. Brewer, James W. and Martha K. Smith. *Emmy Noether: A Tribute to Her Life and Work*. New York: Dekker, 1981.

Peano:

120. Ashurst, F. Gareth. *Founders of Modern Mathematics*. London: Frederick Muller, 1982: p. 111.
121. Ashurst, p. 110.
122. Ashurst, pp. 108, 111.
123. Ashurst, p. 112.
124. Ashurst, pp. 99-113.
125. Kennedy, Hubert C. *Peano: The Life and Works of Giuseppe Peano*. Boston: Dordrecht, 1980.

Ramanujan:

126. *The Man Who Loved Numbers*. Videorecording. Written, produced, and directed by Christopher Sykes. An INCA production, in association with WGBH, Boston and Channel 4, Ltd for NOVA. Northbrook, IL: Distributed by Coronet Film and Video, 1988.
127. *The Man Who Loved Numbers*.
128. *The Man Who Loved Numbers*.
129. *The Man Who Loved Numbers*.
130. *The Man Who Loved Numbers*.
131. *The Man Who Loved Numbers*.
132. Ranganathan, S. R. *Ramanujan: The Man and the Mathematician*. London: Asia, 1967.

Rudin:

133. Albers, Donald J., G. L. Alexanderson, and Constance Reid. "An Interview with Mary Ellen Rudin." *College Mathematics Journal* 19 (March 1988):115.
134. Rudin, Mary Ellen. *Lectures on Set Theoretic Topology*. Providence, RI: American Mathematical Society, 1975. (Regional Conference Series in Mathematics, no. 23): pp. 7, 31, 37.
135. Rudin, pp. 1, 44, 113.
136. Albers, Alexanderson, and Reid, "An Interview," p. 121.
137. Albers, Alexanderson, and Reid, "An Interview," p. 124.
138. Albers, Donald J., G. L. Alexanderson, and Constance Reid, eds. *More Mathematical People*. Boston: Harcourt, Brace, Jovanovich, 1990, pp. 282-303.
139. Kenschaft, Patricia C. "Charlotte Angas Scott, 1858- 1931." *College Mathematics Journal* 18 (March 1987):98-110.
140. Eves, Howard. *An Introduction to the History of Mathematics*. 6th ed. Philadelphia: Saunders, 1990: p. 575.

Computer Revolutionaries:
A Guide to the Literature
on Pioneers in Computing

Charles Matthews

SUMMARY. The development of the first electronic digital computers in the 1940's fulfilled a hypothesis made 100 years earlier by Charles Babbage. The technical advances that allow us to manipulate vast amounts of information and data were made possible by the efforts of a few visionaries. This survey attempts to track the work documenting their efforts.

According to the National Academy of Sciences, the computer is responsible for the second industrial revolution. Indeed, the computer has advanced science and society at breakneck speed by giving us the means to process vast amounts of data and information. These new capabilities are not without their attendant problems: the creation of data at rates faster than can be analyzed; the security and privacy of individuals and organizations; and the added complexity that requires us to adapt old habits and techniques to new technologies. But for most of us, the possibilities offered by this technology outweigh the concerns.

The development of the electronic digital computer was a major scientific revolution. Like all revolutions, it has been driven by a handful of people who were able to articulate a personal vision not

Charlie Matthews received his BA from Fairfield University and MSLS from the University of Kentucky. He currently manages the Technical Information Center for the Apollo Systems Division of Hewlett-Packard Company in Chelmsford, MA. Prior to this, he was R&D Librarian for Apollo Computer Inc. and previously held positions as Technical Librarian for Digital Equipment Corporation in Nashua and Salem, NH.

43

widely shared, or even tolerated, in their time. We are still in the midst of the revolution. As the impact of the computer on our lives continues to grow, it becomes useful to examine those whose brilliance and perseverance brought new possibilities, so that we might improve upon their legacy.

ELECTRONIC DIGITAL COMPUTERS

This literature survey will focus on those who have made a major contribution to the development of the electronic digital computer, which is the basis for the vast majority of computers in use today. This computer has four main characteristics. It is *automatic* in that it works by itself with the aid of a set of coded instructions known as a program that is stored in the computer. It is *general purpose* in that it is able to perform any job that can be described to it in terms of operations that it is able to understand. It is *electronic* in that it utilizes electronic components that allow it to operate at a high speed. It is *digital* in that it represents the data that it works with in the form of discrete digits.[1]

ORIGINS

Tools to aid humans in mathematical calculations date back to prehistory. Early tools, such as the abacus, were manually operated. Mechanical calculating devices made their appearance in the seventeenth century. The most famous of these were built by the French mathematician and philosopher, Blaise Pascal. In 1643, at the age of nineteen, he built a device that could automatically add and subtract columns of numbers, though it wasn't until the nineteenth century when these machines came into practical use. Other mechanical devices that could control sequences of events — such as operating clocks, music boxes or mechanized looms — were working as early as the sixteenth century. In 1805, Joseph M. Jacquard developed a method to program the operation of silk looms using perforated cards. It took the eclectic British mathematician and astronomer, Charles Babbage (1791-1871), to marry the concepts of mechanical calculation and the mechanical control of a sequence of operations together in a machine he called the Analytical Engine in

1834. Babbage's experiments led to the development of logical principles upon which today's computers are based.

SOURCES OF INFORMATION

Babbage is considered by many as the father of the modern computer. His work greatly influenced the twentieth century theoreticians who pioneered the first operating digital computing devices. This survey will cover sources that are largely biographical, beginning with Babbage and spanning more than 100 years of progress in computer science. Research works by figures that have had a major impact on the field are also included.

TOPICAL WORKS

Calculating Engines

Charles Babbage's work in astronomy first prompted him to devise a calculating machine. Babbage was disturbed by the errors he found in astronomical tables and sought a method to produce them mechanically. His stature as a scientist allowed him to secure funding from Parliament to develop his Difference Engine for this purpose. This machine was never completed, but a portion of it was shown to a few interested people, including the mathematician Ada Augusta King, the Countess of Lovelace and daughter of the poet Lord Byron. Ada collaborated with Babbage in designing and documenting the successor of the Difference Engine: the Analytical Engine. Ada's use of a binary numbering scheme to program Babbage's machines has earned her the title of the world's first programmer. Despite Babbage's broad scientific interests, his work on his engines became a lifelong obsession. Neither of Babbage's engines were completed—but they demonstrated to future generations of scientists that a mechanical device was capable of solving complex mathematical equations.

Monographs

Babbage, Charles. *Passages from the life of a philosopher*. 1864. Reprinted, New York: Augustus M. Kelly, 1969.

Babbage, Henry. *Babbage's calculating machines*. 1889. Reprinted with an introduction by Allan G. Bromley in the Charles Babbage Institute Reprint Series for the History of Computing vol. 2, Cambridge, MA: MIT Press, 1984.

Baum, Joan. *The calculating passion of Ada Byron*. Hamden, CT: Archon Book, 1986.

Buxton, H. W. *Memoir of the life and labours of the late Charles Babbage Esq*, F.R.S. The Charles Babbage Institute Reprint Series for the History of Computing, vol. 13, Cambridge, MA: MIT Press, 1987.

Dubbey, J.M. *The mathematical work of Charles Babbage*. Cambridge: Cambridge University Press, 1978.

Elwin, Malcolm. *Lord Byron's family: Annabella, Ada and Augusta, 1816-1824*. London: John Murray, 1975.

Franksen, Ole I. *Mr. Babbage's secret: the tale of a cypher and APL*. Englewood Cliffs, NJ: Prentice-Hall, 1984. Examines Babbage's renown as a cryptographer.

Goldstine, Herman H. *A history of numerical analysis from the 16th through the 19th century*. New York: Springer-Verlag, 1977.

Hyman, Anthony. *Charles Babbage: pioneer of the computer*. Princeton, NJ: Princeton University Press, 1982. Documents Babbage's life and work in great detail.

Hyman, Anthony, ed. *Science and reform* (selected works of Charles Babbage). New York: Cambridge University Press, 1989.

Lindgren, Michael, trans. by Craig G McKay. *Glory and failure: the difference engines of Johann Muller, Charles Babbage, and Georg and Edvard Scheutz*. MIT Press Series in the History of Computing, Cambridge, MA: MIT Press, 1990.

Merzbach, Uta. *Georg Scheutz and the first printing calculator*. Smithsonian Studies in History and Technology, n.36, Washington, DC: Smithsonian Institution Press, 1977.

Moore, Doris Langley. *Ada, Countess of Lovelace: Byron's legitimate daughter*. New York: Harper & Row, 1977.

Morrison, Philip and Emily Morrison, eds. *Charles Babbage and his calculating engines, including "Passages from the life of a philosopher" and "Sketch of the Analytical Engine invented by Charles Babbage by L.F. Menabrea" and "Notes upon the memoirs by the translator, Ada Augusta, Countess of Lovelace." New York: Dover, 1961.

Moseley, Maboth. *Irascible genius: a life of Charles Babbage, inventor.* London: Hutchinson, 1964.

Stein, Dorothy. *Ada: a life and a legacy.* MIT Press Series in the History of Computing, Cambridge, MA: MIT Press, 1985.

Electromechanical Solutions

By the late nineteenth century, mechanical engineering had advanced to the point where it was possible to build large-scale calculating machines capable of direct multiplication and division as well as addition and subtraction. Perhaps the most famous of these were punched-card tabulating machines designed by Herman Hollerith, a first-generation American of German descent. One of Hollerith's electromechanical machines was used to tabulate the 1890 United States Census. Geoffrey Austrian's biography provides a detailed account of one of the great computing pioneers, whom some consider to have built the first large-scale functional computer. By the turn of the century, scores of companies were manufacturing calculating devices—most of them mechanical adding machines. In 1911, Hollerith's Tabulating Machine Company became, through a series of mergers, part of a company that would later become IBM.

Further advances in calculator design took place prior to World War II. At Bell Labs, George Stibitz led a team responsible for the building of the first electromechanical relay-based binary calculators. In Germany, Konrad Zuse made remarkable progress in designing the most advanced electromechanical computers of the time, unaware of parallel developments in the United States and Britain. Fortunately, the Nazi government took little notice of Zuse's work and he was able to flee to Switzerland toward the end of the war with a working prototype. Zuse continued to design and manufacture computers until his company was absorbed by Siemens AG in 1969. In 1939, IBM's Thomas Watson, Sr. and Harvard mathematician Howard Aiken joined in a shaky alliance to develop the Automated

Sequence-Controlled Calculator, later known as the MARK I. The MARK I and its successors represented the state of the art in calculating machinery. But their success was short-lived due to the advent of electronic machines. This period is thoroughly documented in Cerruzi's *Reckoners*.

Monographs

Austrian, Geoffrey D. *Herman Hollerith: forgotten giant of information processing*. New York: Columbia University Press, 1982.

Ceruzzi, Paul. *Reckoners: the prehistory of the digital computer, from relays to the stored program concept, 1935-1945*, Westport, CT: Greenwood Press, 1983.

Cohen, I. Bernard. *Howard H. Aiken, computer pioneer*. Cambridge, MA: MIT Press, forthcoming.

d'Ocagne, Maurice. *Le calcul simplifie : graphical and mechanical methods for simplifying calculation*. 1928. Translated from the French by J. Howlett and M.R. Williams, 3rd ed., in the Charles Babbage Institute Reprint Series for the History of Computing, vol. 11, Cambridge, MA: MIT Press, 1986.

Eckert, Wallace J. *Punched card methods in scientific computation*. 1940. Reprinted with an introduction by J. C. McPherson in the Charles Babbage Institute Reprint Series for the History of Computing, vol. 5, Cambridge, MA: MIT Press, 1984.

The Harvard Computation Laboratory. *A manual of operation for the Automated Sequence Controlled Calculator*. 1946. Reprinted with an introduction by I. Bernard Cohen in the Charles Babbage Institute Reprint Series for the History of Computing, vol. 8, Cambridge, MA: MIT Press, 1985.

Horsburgh, Ellice M., ed. *Handbook of the Napier Tercentenary Celebration or Modern instruments and methods of calculation*. 1914. Reprinted with an introduction by M. R. Williams in the Charles Babbage Institute Reprint Series for the History of Computing, vol. 3, Cambridge, MA: MIT Press, 1984.

Truesdell, Leon E. *The development of punch card tabulation in the Bureau of the Census, 1890-1940*. Washington, DC: Department of Commerce, Bureau of the Census, 1965.

Zuse, Konrad. *Der Computer, mein Lebenswerk*. Mit Geleitworten von F.L. Bauer und H. Zemanek, 2nd ed. Berlin: Springer, 1986. In German. Not translated into English to my knowledge.

Wartime Resolve

Development efforts accelerated during and after World War II as a result of massive infusions of funding and talent. By 1943, a team at the University of Pennsylvania's Moore School of Electrical Engineering, led by John Mauchly and J. Presper Eckert, were commissioned by the U.S. Army to devise a machine to quickly calculate gunnery firing tables. Their work resulted in the ENIAC, thought to be the first electronic calculator when it was completed in 1945. ENIAC weighed 30 tons and was 500 times faster than the Mark I. While it was the fastest general purpose calculator of its time, it had no ability to store programs. John von Neumann, of Los Alamos fame, joined the team to work on a successor, the EDVAC, that would overcome this limitation. Von Neumann's method for storing programs in memory would allow machines to automatically control processes, giving birth to the first true computers.[2] However, the honor of being the first computer to run a stored program belonged to the SSEC, which was completed by IBM in 1948.

Technical advances were also taking place in England during this period. Researchers from Manchester and Cambridge Universities had visited the United States during the 1930's and 1940's, including one of computing's greatest theoreticians, Alan Turing. In his landmark 1936 paper "On Computable Numbers," Turing conceptualized the *Turing Machine* that could calculate any computable function. Turing's interest in computing quickly shifted from theoretical to practical during the war, when his talents turned to designing *Colossus*, a machine instrumental in breaking German codes. See Andrew Hodge's book for an outstanding account of Turing's life and work.

The postwar period brought with it a frenzy of interest in computers. Turing went to work for England's National Physical Laboratory on the ACE (Automatic Computing Engine) and its successor, DEUCE. Eckert and Mauchly established the Eckert-Mauchly Computer Corporation, America's first computer manufacturing company. In 1949, Maurice Wilkes' EDSAC began operation at the

University of Cambridge as the world's first fully-electronic stored program computer, while in the United States, government research continued to result in more advanced machines, such as MIT's Whirlwind in 1951.

One other major, if not curious, event occurred in this period. In 1942, John V. Atanasoff and Clifford Berry completed the Atanasoff-Berry Computer (ABC), an electronic computer with digital memory, at Iowa State College (now Iowa State University). The work was never patented, never demonstrated in a completed form, and barely publicized. Its contribution to the field wasn't fully disclosed to the public until the 1966 publication of R. K. Richards' *Electronic Digital Systems*. Yet in 1973, the courts invalidated the ENIAC patent, declaring ABC instead to be the first electronic digital computer — an assertion that is still hotly debated in the literature.

Monographs

Aspray, William. *John von Neumann and the origins of modern computing*. MIT Press Series in the History of Computing, Cambridge, MA: MIT Press, 1990.

Bolter, J. David. *Turing's man: Western culture in the computer age*. Chapel Hill, NC: University of North Carolina Press, Chapel Hill, 1984.

Burks, Alice R. and Arthur W. *The first electronic computer: the Atanasoff story*. University of Michigan Press, Ann Arbor, 1988.

Campbell-Kelly, Martin, and Williams, Michael R., eds. *The Moore School lectures: theory and techniques for design of electronic digital computers*. 1946. Reprinted in the Charles Babbage Institute Reprint Series for the History of Computing, vol. 9, Cambridge, MA: MIT Press, 1985.

Deavours, Cipher A., David Kahn, Louis Kruh, Greg Mellen, and Brian Winkel, eds. *Cryptography: yesterday, today, and tomorrow*. Artech House, Dedham, MA, 1987.

Diebold, John. *The world of the computer*. New York: Random House, 1973.

Engineering Research Associates Staff. *High-speed computing devices*. 1950. Reprinted with an introduction by Arnold A. Cohen in the Charles Babbage Institute Reprint Series for the History of Computing, vol. 4, Cambridge, MA: The MIT Press, 1984.

Forrester, Jay W. *Collected papers of Jay W. Forrester*. Cambridge, MA: MIT Press, 1975. Forrester invented magnetic core memory and led the team that developed the first real-time computer, MIT's Whirlwind.

Hartree, Douglas R. *Calculating machines: recent and prospective developments and their impact on mathematical physics* and *Calculating instruments and machines*. 1949. Reprinted with an introduction by Maurice Wilkes in the Charles Babbage Institute Reprint Series for the History of Computing, vol. 6, Cambridge, MA: MIT Press, 1984.

Heims, Steve J. *John von Neumann and Norbert Wiener: from mathematics to the technologies of life and death*. Cambridge, MA: MIT Press, 1980.

Hodges, Andrew. *Alan Turing: the enigma*. New York: Simon & Schuster, 1983.

Kahn, David. *The codebreakers: the story of secret writing*. New York: Macmillan, 1967.

Lavington, Simon. *Early British computers: the story of vintage computers & the people who built them*, Manchester: Manchester University Press; Bedford, MA: Digital Press, 1980.

Lukoff, Herman. *From dits to bits; a personal history of the electronic computer*. Portland, OR: Robotics Press, 1979. Not indexed.

Mollenhoff, Clark R. *Atanasoff: forgotten father of the computer*. Ames, IA: Iowa State University Press, 1988.

Redmond, Kent C. and Thomas Malcolm Smith. *Project Whirlwind; the history of a pioneer computer*. Bedford, MA: Digital Press, 1980.

Richards, R. K. *Electronic digital systems*. New York: Wiley, 1966.

Stern, Nancy. *From ENIAC to UNIVAC: an appraisal of the Eckert-Mauchly computers*. Bedford, MA: Digital Press, 1981.

Turing, Alan M. *A.M. Turing's ACE report of 1946 and other papers*. Edited by B.E. Carpenter and R.W. Doran in the Charles Babbage Institute Reprint Series for the History of Computing, vol. 10, Cambridge, MA: MIT Press, 1986.

Turing, Sarah. *A.M. Turing*. Cambridge: Heffer & Sons, 1959.

von Neumann, John. *Collected works*. 6 vols., Elmsford, NY: Pergamon Press, 1961-1963.

von Neumann, John. *The computer and the brain*. New Haven: Yale University Press, 1958. von Neumann's final essays written prior to his death in 1957.

von Neumann, John. *Papers of John von Neumann on computers and computer theory*. Reprinted with an introduction by Arthur W. Burks in the Charles Babbage Institute Reprint Series for the History of Computing, vol. 12, Cambridge, MA: MIT Press, 1986.

Wilkes, Maurice V. *Memoirs of a computer pioneer*. MIT Press Series in the History of Computing, Cambridge, MA: MIT Press, 1985.

Winterbotham, Frederick William. *The ultra secret*. New York: Harper & Row, 1974.

Wulforst, Harry. *Breakthrough to the computer age*. New York: Charles Scribner's Sons, 1982.

Conference Proceedings

Bowden, B. Vivien (Lord Bowden), ed. *Faster than thought: a symposium on digital computing machines*. London: Pitman Publishing, 1953.

Glimm, James and John Impagliazzo, eds. *The legacy of John von Neumann*. Proceedings of Symposia in Pure Mathematics, vol. 50, Providence, RI: American Mathematical Society, 1990.

The Harvard Computation Laboratory. *Proceedings of a symposium on large-scale digital calculating machinery*. 1947. Reprinted with an introduction by William Aspray in the Charles Babbage Institute reprint series for the history of computing, vol. 7, Cambridge, MA: MIT Press, 1985.

Herken, R., ed. *The Universal Turing Machine. A half-century survey*, (Berlin, 1988). New York: Oxford University Press, 1988.

Williams, Michael R. and Martin Campbell-Kelly, eds. *The early British computer conferences*. Reprinted in the Charles Babbage Institute Reprint Series for the History of Computing, vol. 14, Cambridge, MA: MIT Press, 1989.

Birth of an Industry

After the war some of those who were successful at designing computers took to commercializing them. By the close of the 1940's there were only two companies doing this: Eckert-Mauchly Computer Corporation and IBM. The former had an abundance of technical expertise, while the latter knew how to sell that expertise. Eckert and Mauchly quickly ran into financial difficulties and sold out to Remington Rand in 1950. In March 1951, Remington Rand delivered its first computer, the UNIVAC I, to the US Bureau of the Census. In 1952, UNIVAC was used to help CBS predict Eisenhower's victory over Stevenson in the presidential election.

It was the beginning of the golden age of the mainframe. While UNIVAC was capturing the public's imagination, IBM was capturing the market. Though Thomas Watson, Sr. was instrumental in introducing computers to IBM's product line, it was Thomas, Jr. who guided IBM into its dominating role in the market. Other successful electronics giants attempted to break into the market in the 1950's but none were able to master that difficult combination of technical prowess and marketing know-how. The mainframe computer industry soon became known as Snow White (IBM) and the seven dwarfs (Burroughs, CDC, GE, Honeywell, NCR, RCA, and Univac), and later as IBM and the BUNCH (Burroughs, Univac, NCR, CDC, and Honeywell). IBM's dominance is reflected in the literature, a selection of which appears below. This fascination ranges from unbridled adulation, such as in Belden's *The Lengthening Shadow: The Life of Thomas J. Watson*, to the outright contempt found in DeLamarter's *Big Blue: IBM's Use and Abuse of Power*. Two of the most informative accounts on IBM are Fishman's *The Computer Establishment*, which covers the industry from the 1950's through the 1970's when IBM was the computer establishment, and Sobel's *I.B.M.: Colossus in Transition* written by a professor of history. Some good personal memoirs of this era include: Bashe's *IBM's Early Computers*; Grosch's *Computer: Bit Slices from a Life*; Lundstrom's *A Few Good Men from Univac*; and Thomas Watson, Jr.'s candid *Father, Son & Co.: My Life at IBM and Beyond*.

Monographs

Allyn, Stanley. *My half century with NCR*. New York: McGraw-Hill, 1967.

Bashe, Charles J., Lyle R. Johnson, John H. Palmer, and Emerson W. Pugh. *IBM's early computers*. MIT Press Series in the History of Computing, Cambridge, MA: MIT Press, 1985.

Belden, Thomas and Marva. *The lengthening shadow: the life of Thomas J. Watson*. Boston: Little, Brown, 1962.

Buchholz, Werner, ed. *Planning a computer system; Project Stretch*. New York: McGraw-Hill, 1962.

DeLamarter, Richard Thomas. *Big Blue: IBM's use and abuse of power*. New York: Dodd, Mead & Company, 1986.

Engelbourg, Saul. *International Business Machines; a business history*. New York: Arno Press, 1976.

Fisher, Franklin M., James W. McKie, and Richard B. Mancke. *IBM and the U.S. data processing industry: an economic history*. New York: Praeger, 1983.

Fishman, Katherine Davis. *The computer establishment*. New York: McGraw-Hill, 1981.

Forman, Richard L. *Fulfilling the computer's promise: the history of Informatics, 1962-1982*. Informatics General Corporation, 1985.

Foy, Nancy. *The sun never sets on IBM: the culture and folklore of IBM World Trade*, New York: William Morrow, 1975.

Glass, Robert L. *Computing catastrophes*. Seattle, WA: Computing Trends, 1983. Chronicles the big failures in the industry: GE, RCA, Xerox, Singer, Viatron as well as some of IBM's missteps. Primarily a reprinting of Datamation articles from the last 20 years.

Glass, Robert L. *The universal elixir and other computing projects which failed*. Seattle, WA: Computing Trends, 1977.

Grosch, Herbert R. *Computer: bit slices from a life*. Lancaster, PA: Underwood-Miller, 1990.

Hendry, John. *Innovating for failure: government policy and the early British computer industry*. MIT Press Series in the History of Computing, Cambridge, MA: MIT Press, 1990.

Lundstrom, David E. *A few good men from Univac*. MIT Press

Series in the History of Computing, Cambridge, MA: MIT Press, 1987.

Malik, Rex. *And tomorrow the world? Inside IBM*. London: Millington, Ltd., 1975.

Norris, William C. *New frontiers for business leadership*. Minneapolis: Dorn Books, 1983. Some thoughts from the founder of Control Data Corporation.

Pugh, Emerson W. *Memories that shaped an industry: decisions leading to the IBM System/360*. MIT Press Series in the History of Computing, Cambridge, MA: MIT Press, 1984.

Rifkin, Glen and George Harrar. *The ultimate entrepreneur: the story of Ken Olsen and Digital Equipment Corporation*. Chicago: Contemporary Books, 1988. The unauthorized biography of the man whom, in 1986, Fortune magazine called ". . . the most successful entrepreneur in the history of American business. . . ."

Rodgers, F. G. "Buck" with Robert L. Shook. *The IBM way: insights into the world's most successful marketing organization*. New York: Harper & Row, 1986.

Rodgers, William. *Think: A biography of the Watsons and I.B.M.* New York: Stein & Day, 1969.

Sobel, Robert. *I.B.M.: Colossus in transition*. New York: Times Books, 1981.

Wang, Dr. An with Eugene Linden. *Lessons: an autobiography*. Reading, MA: Addison-Wesley, 1986.

Watson, Thomas J. Jr. *A business and its beliefs: the ideas that helped build IBM*, New York: McGraw-Hill, 1963.

Watson, Thomas J. Jr. *Father, Son & Co.: my life at IBM and beyond*, Bantam Books, 1990.

Orders of Magnitude

The tremendous size and power consumption of the first generation of vacuum-tube-based computers proved to be a limiting factor. In 1947 the trio of Walter Brattain, William Shockley and John Bardeen invented the transistor — the first semiconductor device — at Bell Labs, a feat that won them the Nobel Prize in Physics in 1956. Manufacturing transistors in production quantity proved to be a problem for the next ten years until the introduction of the Philco 2000—the first transistor-based computer—in 1957.

While computer design flourished on the east coast, Frederick Terman, a Stanford professor, was instrumental in establishing a microelectronics industry on the west coast. Terman urged his students to remain in California and helped to establish the Stanford Industrial Park, where graduates could start up their own companies.

Shockley returned to Palo Alto in 1954 where he founded Shockley Semiconductor Laboratories, the first semiconductor company in the "Silicon Valley," as a subsidiary of Beckman Instruments. Several engineers later left Shockley to join the new semiconductor unit of Fairchild Camera and Instruments. Robert Noyce, a physicist, joined them at Fairchild. Noyce saw this as the beginning of the entrepreneurial spirit in the valley, when salaries soon gave way to stock-equity in start-ups for young engineers.[3]

The 1957 launching of Sputnik spurred the Department of Defense to fund dozens of projects in solid-state electronics. The idea of placing many transistors on a single chip came independently to Jack Kilby at Texas Instruments and Fairchild's Bob Noyce. Both filed patents at the same time for a semiconductor integrated circuit. Noyce won the patent in court ten years later but both are known as co-inventors of the integrated circuit. Braun's *Revolution in Miniature* provides one of the best non-technical histories of semiconductor electronics. Two books that thoroughly cover solid state electronics and the integrated circuit are Queisser's *The Conquest of the Microchip* and Reid's *The Chip*.

Monographs

Braun, Ernest and Stuart MacDonald. *Revolution in miniature: the history and impact of semiconductor electronics*, 2nd ed. Cambridge: Cambridge University Press, 1982.

Bylinsky, Gene. *The innovation millionaires: how they succeed*. New York: Scribner's 1976. Includes a chapter on Intel founders Bob Noyce and Gordon Moore.

Evans, Christopher. *The making of the micro*. New York: Van Nostrand Reinhold, 1981.

Handel, Samuel. *The electronic revolution*. Baltimore: Penguin, 1967. A history of electronics up to the invention of the transistor.

Hansen, Dirk. *The new alchemists: Silicon Valley and the microelectronics revolution*. Boston: Little, Brown, 1983.

Kikuchi, Makoto. *Japanese electronics — a worm's eye view of its evolution*. Tokyo: Simul Press, 1983.

Mahon, Thomas. *Charged bodies; people, power, and paradox in Silicon Valley*. New York: New American Library, 1985.

Malone, Michael S. *The big score: the billion dollar story of Silicon Valley*. New York: Doubleday, 1985.

Millman, S., ed. *Physical Sciences 1925-1980*. AT&T Bell Labs, 1983. Part of the history of AT&T Bell Labs, covering development work on semiconductors and transistors.

Queisser, Hans. *The conquest of the microchip*. Cambridge, MA: Harvard University Press, 1990.

Reid, T.R. *The chip: how two Americans invented the microchip and launched a revolution*. New York: Simon & Schuster, 1984.

Rogers, Everett M. and Judith K. Larsen. *Silicon valley fever: growth of high-technology culture*. New York: Basic Books, 1984.

Sze, Simon M., ed. *Semiconductor Devices: Pioneering Papers*. Singapore: World Scientific, 1990.

Shockley, William. *Electrons and holes in semiconductors, with applications to transistor electronics*. New York: Van Nostrand, 1950. Reprint. Huntington, NY: Krieger,

Shifting Bits

Advances in microelectronics were paralleled by other refinements. Instructing a computer to perform tasks proved laborious for programmers using machine-specific low-level languages. By 1957, IBM had released FORTRAN, a high-level language devised by John Backus that allowed programmers to instruct computers using mathematical expressions rather than machine or assembly code. In 1959, Grace Murray Hopper was instrumental in creating a standard programming language — COBOL — that utilized natural language expressions. The emergence of high-level languages, along with other developments such as timesharing and the introduction of the cathode ray tube terminal, made computers more accessible to a growing number of people.

The hardware and software developments of the period following World War II are best represented in the recollections of their participants in the *ACM Turing Award Lectures: The First Twenty Years: 1966-1985*. This ongoing series covers history, reminiscences, theory, and visions of the future. Lammers' *Programmers at Work* interviews some of the world's renowned programmers to explain how they work, while Kidder's critically acclaimed *The Soul of a New Machine* provides a fly-on-the-wall look at the development of a new Data General minicomputer. Stewart Brand's *The Media Lab* includes extensive interviews with MIT researchers who are examining future applications for computers. Finally, Wilkes' 1951 publication, *The Preparation of Programs for an Electronic Digital Computer*, is considered to be the world's first computer programming textbook.

Monographs

ACM Turing award lectures: the first twenty years: 1966-1985. Reading, MA: Addison-Wesley, ACM Press, 1987.

Brand, Stewart. *The media lab: inventing the future*. New York: Viking, 1987.

Carlston, Douglas G. *Software people: inside the computer business*, Englewood Cliffs, NJ: Prentice-Hall, 1986.

Dijkstra, Edsger W. *Selected writings on computing: a personal perspective*. New York: Springer-Verlag, 1982.

Feigenbaum, Edward A. and Pamela A. McCorduck. *The fifth generation: artificial intelligence and Japan's computer challenge to the world*. New York: Addison-Wesley, 1983.

Freiherr, Gregory. *The seeds of artificial intelligence*. Washington, DC: Government Printing Office, 1980.

Hilts, Philip J. *Scientific temperaments: three lives in contemporary science*. New York: Simon & Schuster, 1984. Includes a biography of John McCarthy, one of the inventors of timesharing and "father of AI."

Kidder, Tracy. *The soul of a new machine*. Boston: Little, Brown and Co., 1981.

Kurzweil, Raymond. *The age of intelligent machines*. Cambridge, MA: MIT Press, 1990.

Lammers, Susan. *Programmers at work*. Redmond, WA: Microsoft, 1986.

McCorduck, Pamela. *Machines who think: a personal inquiry into the history and prospects of artificial intelligence*. San Francisco: W.H. Freeman, 1979.

Rheingold, Howard. *Tools for thought; the people and ideas behind the next computer revolution*. New York: Simon & Schuster, 1985.

Rose, Frank. *Into the heart of the mind: an American quest for Artificial Intelligence*. New York: Harper & Row, 1984.

Sammet, Jean E. *Programming languages: history and fundamentals*. Englewood Cliffs, NJ: Prentice-Hall, 1969.

Steele, Guy L. Jr. *The Hacker's dictionary: a guide to the world of computer wizards*, New York: Harper & Row, 1983.

Swartzlander, Earl E., Jr. *Computer design development: principal papers*. Rochelle Park, NY: Hayden Book, 1976.

Ulam, Stanislaw. *Science, computers, and people: from the tree of mathematics*. Cambridge, MA: Birkhauser Boston, 1986.

Wilkes, Maurice V.; David J. Wheeler, and Stanley Gill. *The preparation of programs for an electronic digital computer*. 1951. Reprinted with an introduction by Martin Campbell-Kelley in the Charles Babbage Institute Reprint Series for the History of Computing, vol. 1, Cambridge, MA: MIT Press, 1984.

Conference Proceedings

Wexelblat, Richard L., ed. *Proceedings: History of Programming Languages Conference* (June 1-3, 1978). New York: Academic Press, 1981. Preprints published as "ACM History of Programming Languages Conference," *ACM Sigplan Notices*, 13, no.8 (August, 1978).

For Home or Office

Thirty years after its invention, the digital computer was still sequestered from its users in climate-controlled, raised floor chambers. Programmers and system administrators formed a priesthood that interceded with it on behalf of the users needs. At the same time, there was a new generation of users with an overwhelming urge to explore the mysteries of the machine. These were mostly students, involved in model railroading, rocketry, or radio electronics clubs with others of like desire. They were determined to explore computers, even if it meant staying up all night with a batch processing machine, or later, as integrated circuits became a common commodity, building their own.

The first personal computer was produced by Ed Roberts, founder of Micro Instrumentation and Telemetry Systems, a company that built guidance systems for model rockets, but later spawned the PC industry with the introduction of the MITS Altair 8800 in 1975. Two years later the Apple II appeared, followed in 1981 by the IBM Personal Computer. The PC was literally a revolution within a revolution — making 20-year-olds into millionaires and putting computers in the hands of the millions.

The PC, and later the workstation, for the first time put computers on the desks of office workers. The computer was liberated, much to the chagrin of many an MIS manager. Apple Computer, and its co-founder Steve Jobs, dominate the literature of this period. Butcher's *Accidental Millionaire* relates the story of Steve Job's tenure at Apple in fascinating detail. Frank Rose's *West of Eden* and John Sculley's *Odyssey* offer disparate accounts of what went on at Apple's changing of the guard. Freiberger's *Fire in the Valley*, Mims' *Siliconnections* and Moritz's *The Little Kingdom* all offer insights into the very early days of personal computers. Littman's *Once Upon a Time in ComputerLand* is an example of a Silicon Valley success story gone awry. Some insight into the development of workstations is found in Goldberg's *A history of personal workstations* and Hall and Barry's *Sunburst*.

Monographs

Butcher, Lee. *Accidental millionaire: the rise and fall of Steve Jobs at Apple Computer*. New York: Paragon House Publishers, 1988.
Carlston, Douglas G. *Software people; an insider's look at the personal computer software industry*. New York: Simon & Schuster, 1985.

Chposky, James and Ted Leonsis. *Blue magic; the people, power, and politics behind the IBM personal computer*. New York: Facts on File, 1988.

Freiberger, Paul and Michael Swaine. *Fire in the valley: the making of the personal computer*. Berkeley, CA: Osborne/McGraw-Hill, 1984.

Garr, Doug. *Woz: the prodigal son of the silicon valley*. New York: Avon Books, 1984.

Gasse'e, Jean-Louis. *The third apple: personal computers & the cultural revolution*. San Diego: Harcourt Brace Jovanovich, 1987.

Hall, Mark and Barry, John. *Sunburst: the ascent of Sun Microsystems*. Chicago: Contemporary Books, 1990.

Levering, Robert, Michael Katz, and Milton Moskowitz. *The computer entrepreneurs: who's making it big and how and America's upstart industry*. New York: New American Library, 1984.

Levy, Steven. *Hackers: heroes of the computer revolution*, Garden City, NY: Anchor/Doubleday, 1984.

Littman, Jonathan. *Once upon a time in ComputerLand: the amazing billion-dollar tale of Bill Millard*. Revised and expanded edition. New York: Simon & Schuster, 1990.

Mims, Forrest M. III. *Siliconnections: coming of age in the electronic era*. New York: McGraw-Hill, 1986.

Moritz, Michael. *The little kingdom: the private story of Apple Computer*. New York: William Morrow, 1984.

Nelson, Ted. *Computer lib/dream machines*. Chicago: Hugo's Book Service, 1974.

Osborne, Adam, and Dvorak, John. *Hypergrowth: the rise and fall of Osborne Computer Corporation*. Berkeley, CA: Idthekkethan Publishing, 1984.

Osborne, Adam. *Running wild: the next industrial revolution*. Berkeley, CA: Osborne/McGraw-Hill, 1979.

Rose, Frank. *West of Eden: the end of innocence at Apple Computer*. New York: Viking, 1989.

Sculley, John with Byrne, John A. Odyssey: *Pepsi to Apple . . . A journey of adventure, ideas, and the future*. Harper & Row, New York, 1987.

Smith, Douglas K. and Robert D. Alexander. *Fumbling the future:*

how Xerox invented, then ignored, the first personal computer.
New York: William Morrow, 1988.

Young, Jeffrey S. *Steve Jobs: the journey is the reward.* Glenview,
IL: Scott, Foresman, 1988.

Conference Proceedings

Goldberg, Adele, ed. *A history of personal workstations.* Reading,
MA: Addison-Wesley, ACM Press, 1988. Includes papers and
photographs from the ACM Conference on the History of Personal Workstations.

GENERAL WORKS

There are some superb comprehensive works that provide some
continuity to the contributions made to computing to date. Aspray's
collection — *Computing before Computers* — surveys the technology
prior to 1945. Bernstein's *The Analytical Engine* is a good popular
history of computing. Augarten's *Bit by Bit* and Eames's *A Computer Perspective* are both heavily illustrated. Goldstine's *The
Computer from Pascal to von Neumann* is a classic combination of
history and personal memoirs, written by a man who worked
closely with von Neumann. Moreau's *The Computer Comes of Age*
is a well-researched and organized work, written by IBM's former
manager of scientific development in France, who covers the field
up to 1964. Four books that focus directly on people, not the machines, are: Caddes's *Portraits of Success*, which is primarily a
collection of high-quality photographs of 65 pioneers; Shurkin's
Engines of the Mind; Slater's *Portraits in Silicon*, which also covers
many recent pioneers in software; and Zientara's *The History of
Computing*. Finally, a must for any computing history collection is
Brian Randell's *The Origins of Digital Computers*, a collection of
selected papers by pioneers from Babbage to Wilkes with introduc-
tions to each and extensive references and indexing.

Monographs

Angus, Anne Denny. *The computer people*. London: Faber, 1970.

Ashurst, F. Gareth. *Pioneers of computing*. London: F. Muller, 1983.

Aspray, William F., ed. *Computing before computers*. Ames, IA: Iowa State University Press, 1990.

Augarten, Stan. *Bit by bit: an illustrated history of computers*. New York: Ticknor & Fields, 1984.

Bernstein, Jeremy. *The analytical engine: computers – past, present, and future*. Newly revised, New York: William Morrow, 1981.

Caddes, Carolyn. *Portraits of success; impressions of Silicon Valley pioneers*, Palo Alto, CA: Tioga Publishing; Los Altos, CA: Distributed by William Kaufmann, 1986.

Eames, Charles and Ray Eames. *A computer perspective*. Cambridge, MA: Harvard University Press, 1973.

Giarratano, Joseph C. *Foundations of computer technology*. Indianapolis, IN: Howard W. Sams., 1982.

Goldstine, Herman. *The Computer from Pascal to von Neumann*. Princeton, NJ: Princeton University Press, 1972.

Hollingdale, S.H., and G.C. Toothill. *Electronic computers*. Baltimore: Penguin Books, 1976.

Moreau, Rene. *The computer comes of age; the people, the hardware, and the software*. Translated by J. Howlett. MIT Press Series in the History of Computing, Cambridge, MA: MIT Press, 1984. Originally published as: *Ainsi naquit l'informatique*.

Phillips, E. *Studies in the history of mathematics*. Washington, DC: Mathematical Association of America, 1987.

Pylyshyn, Zenon W. *Perspectives on the computer revolution*. Englewood Cliffs, NJ: Prentice-Hall, 1970. Contains historical background.

Randell, Brian, ed. *The origins of digital computers: selected papers*. 3rd ed., Berlin: Springer-Verlag, 1982.

Rashid, Rick, ed. *CMU Computer Science twenty-fifth anniversary commemorative*. Reading, MA: Addison-Wesley, ACM Press, 1990.

Ritchie, David. *The computer pioneers; the making of the modern computer*. New York: Simon & Schuster, 1986.

Rosenberg, Jerry M. *The computer prophets*. New York: Macmillan, 1969.

Rowe, David and John McCleary. *History of modern mathematics*. Academic Press, 1989.

Shurkin, Joel. *Engines of the mind: a history of the computer*. New York: W.W. Norton, 1984.

Slater, Robert. *Portraits in silicon*. Cambridge, MA: MIT Press, 1987.

Wildes, Karl L., and Nilo A. Lindgren. *A century of electrical engineering and computer science at MIT, 1882-1982*. Cambridge: MIT Press, 1985.

Williams, Michael R. *A history of computing technology*. Englewood Cliffs, NJ: Prentice-Hall, 1985.

Zientara, Marguerite. *The history of computing*. Framingham, MA: CW Communications, 1981.

Conference Proceedings

Metropolis, N., J. Howlett, and G.-C. Rota, eds. *A history of computing in the twentieth century*, Academic Press, New York, 1980. A collection of essays, mostly based on papers given at the International Research Conference on the History of Computing, Los Alamos Scientific Lab, June 1976. Includes authoritative history of ENIAC by Arthur Burks and reminiscences by Eckert and Mauchly.

Nash, Stephen G., ed. *A history of scientific computing*. Reading, MA: Addison-Wesley, ACM Press, 1990. Based on papers presented at the Conference on the History of Scientific and Numeric Computation, held in Princeton, NJ, 1987.

A quarter century of IFIP: the IFIP silver summary. Proceedings of the 25th anniversary celebration of the IFIP (Munich, March 27, 1985). New York: Elsevier North-Holland, 1986. Contains reminiscences, technical papers, biographies, histories, and photographs from the conference honoring the establishment of the International Federation for Information Processing.

TITLES IN SERIES

ACM Press Anthology Series (Addison-Wesley)

ACM Turing award lectures: the first twenty years: 1966-1985. 1987.

Rashid, Rick, ed. *CMU Computer Science twenty-fifth anniversary commemorative.* 1990.

ACM Press History Series (Addison-Wesley)

Goldberg, Adele, ed. *A history of personal workstations.* 1988.

Blum, Bruce, ed. *A history of medical informatics.* 1990.

Nash, Stephen, ed. *A history of scientific computing.* 1990.

Charles Babbage Institute Reprint Series for the History of Computing (MIT Press)

Wilkes, Maurice V.; David J. Wheeler, and Stanley Gill. *The preparation of programs for an electronic digital computer.* 1951. vol.1, 1984.

Babbage, Henry. *Babbage's calculating machines.* 1889. vol. 2, 1984.

Horsburgh, Ellice M., ed. *Handbook of the Napier Tercentenary Celebration or Modern instruments and methods of calculation.* 1914. vol. 3, 1984.

Engineering Research Associates Staff. *High-speed computing devices.* 1950. vol. 4, 1984.

Eckert, Wallace J. *Punched card methods in scientific computation.* 1940. vol. 5, 1984.

Hartree, Douglas R. *Calculating machines: recent and prospective developments and their impact on mathematical physics and Calculating instruments and machines.* 1949. vol. 6, 1984.

The Harvard Computation Laboratory. *Proceedings of a symposium on large-scale digital calculating machinery.* 1947. vol. 7, 1985.

The Harvard Computation Laboratory. *A manual of operation for the Automated Sequence Controlled Calculator.* 1946. vol. 8, 1985.

Campbell-Kelly, M. and Williams, M., eds. *The Moore School lec-*

tures: theory and techniques for design of electronic digital computers. 1946. vol. 9, 1985.

Turing, Alan M. A.M. *Turing's ACE report of 1946 and other papers*. vol. 10, 1986.

d'Ocagne, Maurice. *Le calcul simplifie': graphical and mechanical methods for simplifying calculation*. 1928. vol. 11, 1986.

von Neumann, John. *Papers of John von Neumann on computers and computer theory*. vol. 12, 1986.

Buxton, H. W. *Memoir of the life and labours of the late Charles Babbage Esq, F.R.S.* vol. 13, 1987.

Williams, Michael R. and Martin Campbell-Kelly, eds. *The early British computer conferences*. vol. 14, 1989.

Napier, John. *Rabdology or calculation with rods in two books with an appendix on the high-speed promptuary for multiplication and one book on location arithmetic*. vol. 15, 1990.

MIT Press Series in the History of Computing

Moreau, Rene. *The computer comes of age; the people, the hardware, and the software*. 1984.

Pugh, Emerson W. *Memories that shaped an industry: decisions leading to the IBM System/360*. 1984.

Bashe, Charles J., Lyle R. Johnson, John H. Palmer, and Emerson W. Pugh. *IBM's early computers*. 1985.

Stein, Dorothy. *Ada: a life and a legacy*. 1985.

Wilkes, Maurice V. *Memoirs of a computer pioneer*. 1985.

Lundstrom, David E. *A few good men from Univac*. 1987.

Aspray, William. *John von Neumann and the origins of modern computing*. 1990.

Hendry, John. *Innovating for failure: government policy and the early British computer industry*. 1990.

Lindgren, Michael, trans. by Craig G McKay. *Glory and failure: the difference engines of Johann Muller, Charles Babbage, and Georg and Edvard Scheutz*. 1990.

PERIODICALS

ACM Computing Reviews. New York: Association for Computing Machinery. 1960-. Publishes book and non-book reviews monthly. *Computing Reviews* includes a "History of Computing" category that regularly carries reviews on interest. In September 1990 *CR* published its first article ever. The topic was the still-heated debate over the Eckert-Mauchly vs. Atanasoff controversy.[4]

ACM Computing Surveys. Baltimore: Association for Computing Machinery. 1969-. Although its primary function is to survey technical issues in computing, it provides excellent background material and references. Some very early issues covered historical topics. See: Rosen, S. "Electronic Computers. A Historical Survey" 1(1) and Knuth, D. "Von Neumann's First Computer Program" 2(4).

Annals of the History of Computing. New York: American Federation of Information Processing Societies, Springer-Verlag. 1979-. The only serial dedicated to the comprehensive study of the history of computing. Though the emphasis is primarily on computer hardware developments, it remains an essential component to any computer science collection.

BYTE. New York: McGraw-Hill. 1975-. Frequently chronicles the people involved in the growth of personal computing. The "15th Anniversary Summit," *BYTE* 15 (September 1990) reviews the past 15 years of the PC's existence and consults 63 gurus on the prospects for the next 15 years.

Charles Babbage Institute Newsletter: the Center for the History of Information Processing. Minneapolis: Charles Babbage Institute. 1979-. Reports on holdings acquired by the Institute and discusses other CBI events.

Communications of the ACM. New York: Association for Computing Machinery. 1958-. Commonly known as the CACM, it is the members magazine of the ACM. Though it generally carries technical articles and members news, it often covers topics in computer history and usually publishes the annual Turing Award lecture, given by the recipient of the ACM's most prestigious award for accomplishment. A special issue, "Celebrating ACM's

40th Anniversary,'' 30 (October 1987), profiles some of the Grace Murray Hopper Award winners, an award given to computer scientists for outstanding achievements before the age of 30, and reflects on the history and accomplishments of the ACM.

Computer. Long Beach, CA: IEEE Computer Society. 1966-. The IEEE Computer Society members magazine. It focuses on current topics in computing and often publishes the work of contemporary computer pioneers.

The Computer Museum Annual. Boston: The Computer Museum. Each Annual surveys a topic in computer history complete with photographs and illustrations.

The Computer Museum News. Boston: The Computer Museum. A bimonthly covering news on museum events and other topics of interest in computer history.

Datamation. New York: Technical Publishing. 1955-. As one of the oldest computer trade periodicals still in publication, Datamation has devoted much space to computer history and pioneers. After more than 35 years of publication, it remains an important chronicle of the industry.

Dr. Dobb's Journal. Menlo Park., CA: M & T Pub. 1976-. Like BYTE, Dr. Dobbs has been covering personal computing since the beginning. It occasionally takes a retrospective look at the PC. DDJ's "15th Anniversary issue," (January 1991) revisits the early days of the PC industry and reflects upon its future.

IBM alumni directory. San Antonio, TX: EX-IBM Corporation. An irregularly issued directory of ex-IBMers.

IEEE Center for the History of Electrical Engineering Newsletter. New York: Institute of Electrical and Electronics Engineers. 1982-. Reports on archives documenting contributions to electrical and electronic engineering.

IEEE Spectrum. New York: Institute of Electrical and Electronics Engineers. 1964-. The IEEE members magazine. Occasionally devotes issues to computing and/or historical themes.

IEEE Transactions on Computers. 1952-. New York: Institute of Electrical and Electronics Engineers. A technical publication that originated as the *Transactions of the IRE Professional Group on Electronic Computers*, it has published some fine contributions such as Peter Wegner's "Programming Languages: the First 25 Years" (December 1976).

ISIS. Philadelphia: History of Science Society. 1912-. Occasionally publishes articles on the cultural impact of computers.

Technology and Culture. Chicago: University of Chicago Press. 1959-. Occasionally features articles of interest on this topic.

GUIDES TO THE LITERATURE

Aspray, William F. and B. H. Bruemmer, ed. *Guide to the oral history collection of the Charles Babbage Institute*. Charles Babbage Institute Occasional Paper, 1986.

Bruemmer, B. H. and S. Hochheiser. *The high-technology company. A historical research and archival guide*. Charles Babbage Institute, 1986.

Bruemmer, B. H. ed. *Resources for the history of computing. A guide to U.S. and Canadian Records*. Charles Babbage Institute Occasional Paper, 1986.

Cortada, James W. *An annotated bibliography on the history of data processing*. Westport, CT: Greenwood Press, 1990. Lists many institutional histories.

Cortada, James W. *Archives of data processing: A guide to major U.S. collections*. Westport, CT: Greenwood Press, 1990. List major archives both private and public, including the Charles Babbage Institute Collection.

Cortada, James W. *Historical dictionary of data processing*. Westport, CT: Greenwood Press, 1987. Entries cover major developments in the history of data processing.

Cortada, James W. *Historical dictionary of data processing: organizations*, Westport, CT: Greenwood Press, 1987. Covers organizations of historical importance to the data processing industry.

Gjerde, J.A. *Oral history sources in the Charles Babbage Institute*. Charles Babbage Institute Occasional Paper, 1983.

Kidwell, Peggy A. and Juanita Y. Morris. *Smithsonian computer history project: a combined index to oral histories open to readers*. Washington, DC: National Museum of American History, Smithsonian Institution, 1986.

Rider, Robin E. and Henry E. Lowood. *Guide to sources in northern California for history of science and technology*. Berkeley: Office for History of Science and Technology, University of California, 1985. Regional survey that includes Stanford and Silicon Valley archives.

Samuels, Helen W. *Selective guide to the collections*. Cambridge, MA: Institute Archives and Special Collections, The Libraries, Massachusetts Institute of Technology, 1988.

Tebbenhoff, E. H. *Bibliography of writings on the history of computing*. Charles Babbage Institute Occasional Paper, 1983.

Tropp, Henry S. *Preserving the history of computers*. Montvale, NJ: AFIPS, 1972.

RESEARCH TOOLS

Encyclopedias and Directories

Asimov, Isaac. *Asimov's biographical encyclopedia of science and technology*, 2nd ed. New York: Doubleday, 1982.

Belzer, Jack, ed. *The encyclopedia of computer science and technology*. New York: Dekker, 1977.

Hildebrandt, Darlene Myers, ed. *Computing information directory: a comprehensive guide to the computing literature*, 7th ed., Colville, WA: Hildebrandt, Inc., 1990.

Ralston, Anthony, ed. *Encyclopedia of computer science and engineering*, 2nd ed. New York: Van Nostrand Reinhold, 1983.

Subject Headings

ACM Computing Reviews Classification:

K.2 Computing Milieux—History of Computing—People
K.2 Computing Milieux—History of Computing—[individual name]

Library of Congress Subject Headings:

Computer engineers — Biography
Computer industry — History
Computer programmers
Computers — Biography
Computers — History
Electronic data processing — History
Electronic digital computers — History
Industrialists — Biography
Programming languages (electronic computers) — History

Library Call Numbers

Dewey Decimal:

001.64
004.092
338.4
338.7
621.3819

Library of Congress:

HD9696.C62-64
QA76.17
TK7885.2
TK7889

Library of Congress (1990 Draft Expansion):[5]

QA76.2 (individual biographies. Followed by cutter for individual and cutter for author, e.g., B223 for Babbage, Charles; H54 for Hollerith, Herman)
QA76.21 (autobiographies)
QA76.215 (computer history)

INDEXING AND ABSTRACTING SERVICES

Literature on computing pioneers is found across a wide range of indexes and abstracting services. Publications on a specific individual are much easier to locate than those on the generic topic. One service that does a superb job at indexing works on people in computing is the *ACM Guide to Computing Literature*. 1960-. The GCL's category index includes an entry for the History of Computing with further sub-categories for specific individuals or people in general.

In 1987 both the GCL and *Computing Reviews* were incorporated into the American Mathematical Society's MATHSCI database, file 239 on Dialog. MATHSCI now includes GCL from 1981-present and CR from 1982-present. The entire ACM Computing Reviews Classification System is searchable in MATHSCI. To locate literature on computing pioneers in MATHSCI via Dialog, select ''sf = (computing or acm or cr or gcl) and (k.2(f)people)/de.''

ORGANIZATIONS

There are several organizations dedicated to preserving the accomplishments of computer pioneers. The American Federation of Information Processing Societies (AFIPS), a federation of several organizations, including the ACM, has a standing committee for the History of Computing and continues to sponsor the publication *Annals of the History of Computing*. The Computer Museum in Boston, MA has been in operation since 1979. Though its primary focus is collecting artifacts, it also contains photographs and early documents of interest to computer historians. Many computer pioneers have spoken at the Computer Museum's long-standing lecture series. The Smithsonian Institution has also developed a substantial collection of computing artifacts over the years. The Smithsonian's Computer History Project is a major effort to collect oral histories and other documents important to the development of the computer.

Perhaps the most extensive effort at preserving the history of the computer is being conducted at the Charles Babbage Institute for the History of Information Processing at the University of Minnesota, in Minneapolis. The CBI maintains archives, a clearinghouse for information about the location and content of historical materials, and actively sponsors research into the history of computing and computing pioneers. References to many CBI publications have been included in this survey. For the location of other major archives in computing history see Cortada's *Archives of Data-Processing History*.[6]

NOTES

1. G. J. Morris, "Digital computers." In: Anthony Ralston, ed. *Encyclopedia of science and engineering*, 2nd ed., Van Nostrand Reinhold, 1983.

2. Rene Moreau. *The computer comes of age; the people, the hardware, and the software*. Cambridge, MA: MIT Press, 1984, 37-38.

3. Lee Butcher. *Accidental millionaire: the rise and fall of Steve Jobs at Apple Computer*. New York: Paragon House, 1988, 7-8.

4. Saul Rosen, "The origins of modern computing," *ACM Computing Surveys* 31 (1990): 450-462. Includes responses: "Response: Herman H. Goldstine," 463-4; "Response: The Burkses," 464-67; "Response: John V. Atanasoff II," 467-469; "Response: Nancy Stern," 469-70; "Response: Bernard A. Galler," 470-1; "Response: Clark R. Mollenoff," 471-474; "Response: Allan R. Mackintosh," 474-5; "Response: Eric A. Weiss," 475-77; "Saul Rosen Responds," 477-81.

5. Darlene Myers Hildebrandt, ed. *Computing information directory: a comprehensive guide to the computing literature*, 7th ed., Colville, WA: Hildebrandt, Inc., 1990, 421.

6. James W. Cortada. *Archives of data processing: a guide to major U.S. collections*. Westport, CT: Greenwood Press, 1990.

BIBLIOGRAPHY

Augarten, Stan. *Bit by bit: an illustrated history of computers*. New York: Ticknor & Fields, 1984.

Butcher, Lee. *Accidental millionaire: the rise and fall of Steve Jobs at Apple Computer*. New York: Paragon House, 1988.

Ceruzzi, Paul. *Reckoners: the prehistory of the digital computer, from relays to the stored program concept*, 1935-1945, Westport, CT: Greenwood Press, 1983.

Goldstine, Herman. *The Computer from Pascal to von Neumann*. Princeton, NJ: Princeton University Press, 1972.

Hildebrandt, Darlene Myers, ed. *Computing information directory: a comprehensive guide to the computing literature*, 7th ed., Colville, WA: Hildebrandt, Inc., 1990.

Kuhn, Thomas S. *The structure of scientific revolutions*, 2nd ed., revised and enlarged. Chicago: The University of Chicago Press, 1970.

Moreau, Rene. *The computer comes of age; the people, the hardware, and the software*. Cambridge, MA: MIT Press, 1984.

Ralston, Anthony, ed. *Encyclopedia of science and engineering*, 2nd ed., Van Nostrand Reinhold, 1983.

Reid, T.R. *The chip: how two Americans invented the microchip and launched a revolution*. New York: Simon & Schuster, 1984.

Shurkin, Joel. *Engines of the mind: a history of the computer*. New York: W.W. Norton, 1984.

Biographies of Physicists:
An Annotated Bibliography

Donna E. Cromer

The 16 physicists in this section range in time from Michelson, whose experiments paved the way for the paradigm shift from classical physics to quantum mechanics and relativity, through the original developers of quantum mechanics, to those who took the 'new physics' for granted and moved on. Their nationalities are varied; truly, the profession of physicist is an international one. Because of this, many have worked together, and most probably knew or had at least met everyone else on this list. Most listed here have won a Nobel Prize, many were involved in the development of atomic weapons, and many have felt a deep sense of responsibility for peace in the world and have been active on its behalf. Their lives and contributions to science make for fascinating reading. All the materials listed are English language (many have been translated) and while relatively few are in print, most are held by many libraries and easily obtained through interlibrary loan.

BETHE, HANS ALBRECHT
Born Jul 02, 1906 (Strasbourg, Ger)
Nobel Prize in Physics, 1967

With his early schooling and work in Germany and then in England, Bethe moved to Cornell University in 1937, where he taught

Donna E. Cromer is Collection Development/Selection for Physics Librarian at the Centennial Science and Engineering Library, University of New Mexico, Albuquerque, NM 87131-1466.

until 1975. He was involved in the Manhattan Project and spent some time in Los Alamos. Never a specialist, he made contributions in many areas of physics, winning the Nobel Prize for his discovery of the mechanisms of energy production in the sun.

Bernstein, Jeremy. 1980. *Hans Bethe, Prophet of Energy*. New York: Basic Books, Inc. 212 p.

Based on lengthy interviews with Bethe, this book was originally serialized in the *New Yorker*. A wide range of topics is covered, reflecting Bethe's own extensive research. The science is well covered, for the general reader. The first two parts of the book cover his early years in Germany and America, and his involvement in the development of the atomic bomb. The final third is a very detailed discussion of the many sources of energy, including their feasibility, limits, and possible future.

BOHR, NIELS HENRIK DAVID
Born Oct 07, 1885 (Copenhagen)
Died Nov 18, 1962 (Copenhagen)
Nobel Prize in Physics, 1922

One of the greatest physicists and statesmen of the 20th century, Bohr was born in Copenhagen and spent most of his life there. He lived some years in England and America, but returned to Copenhagen to direct the new Institute for Theoretical Physics until the end of his life. The years of World War II were spent in America, where he was involved in the Manhattan Project. One of the originators of quantum physics, Bohr is well known for his theories of the structure of the atom (for which he won the Nobel Prize) and for his views on complementarity, which was originally developed to help understand two seemingly incompatible behaviors, such as the wave-like and particle-like behaviors of light.

Blaedel, Niels. 1988. *Harmony and Unity: The Life of Niels Bohr*. Madison, WI: Science Tech Publishers. Translated by Geoffrey French. 323 p.

Physics and Bohr's contributions to the field are explained, for the most part, in language for the general reader. His ideas of complementarity (which extend beyond physics) are covered in detail, as is one of his last, and very different, works on the relationship between physics and biology. Also considered are Bohr's connections with other physicists, and his impact on politics (primarily his attempts to halt the overdevelopment of nuclear weapons). Many photographs are included.

French, A.P. and P.J. Kennedy. 1985. *Niels Bohr: A Centenary Volume*. Cambridge, MA: Harvard University Press. 403 p.

The science, politics, and philosophy of Bohr are covered by many contributors in this nicely done celebration of Bohr's birth. The level of scientific background necessary for understanding varies greatly. There are numerous photographs and reproductions of original articles.

Moore, Ruth. 1966. *Niels Bohr: The Man, His Science, & the World They Changed*. New York: Alfred A. Knopf. 436 p. (reprinted in paperback in 1985 by MIT Press).

A reverent portrait of a great man, all facets of Bohr's life are covered. The science, including Bohr's contributions to the revolutionary quantum physics is explained in language for the nonphysicist. Bohr's earlier years are described in detail, including a rather dramatic account of his escape from the Nazis and his flight from Denmark to Sweden and England. His involvement in politics in his later years is also discussed.

Rozental, S. 1967. *Niels Bohr: His Life and Work as Seen by His Friends and Colleagues*. Amsterdam: North Holland. 355 p.

Contributions by friends, family, and colleagues describe the various periods of Bohr's life and scientific activity in physics and other fields. Personal recollections conclude the volume.

CURIE, MARIE SKLODOWSKA
Born Nov 07, 1867 (Warsaw)
Died Jul 04, 1934 (Sancellemoze, Fr)
Nobel Prize in Physics, 1903; in Chemistry, 1911

The first woman to win a Nobel Prize and the first person to ever win two, Curie was a remarkable person. Born in Poland, she desired more opportunities and in 1891 left for Paris for advanced study. She spent most of the rest of her life in France, working tirelessly to penetrate the mysteries of the newly discovered radiation. Her Nobel Prize in physics, shared with Henri Becquerel and Pierre Curie, was for research into the phenomenon of radiation. The second Nobel Prize, in Chemistry, was primarily for her discovery of the elements radium and polonium.

Curie, Eve. 1937. *Madame Curie: A Biography*. New York: Doubleday, Doran & Co., Inc. Translated by Vincent Sheean. 412 p.

A romanticized account of Curie's life and work, this was written by her youngest daughter. The earlier years, before fame overtook Curie and her husband are covered in detail. There are extensive quotes from letters and diaries.

Giroud, Françoise. 1986. *Marie Curie: A Life*. New York: Holmes & Meier. Translated by Lydia Davis. 291 p.

Beginning with her early years, this is a personal, not scholarly, account of Curie's life. Concentrating on Curie's nonprofessional life, while not ignoring Curie as a scientist nor her science, this ends with her death from radiation poisoning. Some photographs are included.

Magee, Judith. 1989. *Marie Curie: A Study of American's Use of Marie Curie As a Devoted Wife and Mother, Saintly Scientist, and Healer of Humanity*. M.A. Thesis. Sarah Lawrence College. var p.

This is primarily an account, using the popular press literature of the time, of the Curies' visit to America in 1921 to raise money to buy radium for research. Her early life is also considered to set the context. Not only a study of the popularization of Curie by the media, this thesis also investigates women's role and status in the U.S. and European science profession and existing female support networks for women scientists.

Pflaum, Rosalynd. 1989. *Grand Obsession: Madame Curie and Her World*. New York: Doubleday. 496 p.

Broader in scope than the title implies, this book describes the life of Curie and continues (after her death in 1934) with the lives of her children, ending with the death of Frédéric Joliot-Curie, the husband of Irène, Curie's eldest daughter. The science is covered in detail (for the nonspecialist). Some more controversial aspects of Curie's life are described, painting a complete portrait of a woman.

Reid, Robert. 1974. *Marie Curie*. London: William Collins Sons & Co., Ltd. 347 p.

Curie's personal life and scientific work and achievements are covered. She is portrayed as a real person, and the more controversial aspects of her life and career are not ignored. The beginnings of the suspicion of the harm radium could cause are discussed.

DIRAC, PAUL ADRIEN MAURICE
Born Aug 08, 1902 (Bristol)
Died Oct 20, 1984 (Tallahassee)
Nobel Prize in Physics, 1933

Born in England (although a Swiss citizen until 1919) Dirac spent his youth and professional career there until he moved to Florida in the early 1970's, where he was at Florida State University until his death in 1984. Dirac was a more private person than many other physicists, and virtually unknown outside the physics community. He is best known for his work in quantum physics, reconciling the two seemingly contradictory views of wave mechanics and matrix mechanics. He shared the Nobel Prize for his discovery of new productive forms of atomic theory.

Kragh, Helge. 1990. *Dirac: A Scientific Biography*. Cambridge: Cambridge University Press. 389 p.

Dirac's personal life and his contributions to science are given thorough treatment. As the title implies, his science is covered in depth and requires a good background in physics for complete understanding. Still, since the biographical accounts are concentrated in four chapters, much of the book can be read and enjoyed by the nonphysicist. Dirac's views on the more general and philosophical aspects of physics are also covered.

Kursunoglu, Behram N. and Eugene P. Wigner (Eds.). 1987. *Reminiscences About a Great Physicist: Paul Adrien Maurice Dirac*. Cambridge: Cambridge University Press. 297 p.

Short contributions from Dirac's wife and 23 friends and colleagues create a picture of an extraordinary, yet private and introverted, physicist and man. His personal life and contributions to many fields of science are covered.

Mehra, Jagdish and Helmut Rechenberg. 1982. *The Historical Development of Quantum Theory. Volume 4*. New York: Springer-Verlag. 322 p.

The development of quantum theory between 1925 and 1926 is seen through the life and work of Dirac. His intellectual development from a child through his student days and completion of his dissertation at Cambridge are described. Many details of how he worked, who he worked with, and how he arrived at his conclusions are provided.

Taylor, J.G. 1987. *Tributes to Paul Dirac*. Bristol: Adam Hilger. 123 p.

Personal reminiscences are a nice contrast to the rather dry biographical sketch of the beginning. The second half of the book describes in detail Dirac's scientific contributions.

FERMI, ENRICO
Born Sep 29, 1901 (Rome)
Died Nov 28, 1954 (Chicago)
Nobel Prize in Physics, 1938

Born, raised, and schooled in Italy, Fermi was one of a small number of physicists with a genius for both theoretical and experimental work. Because his wife was Jewish, he and his family left Italy for the United States in 1938. The discovery of fission had just been announced, and by December 1942, Fermi, with the help of many others, had achieved the first sustained nuclear reaction. Fermi spent the rest of the years of World War II at Los Alamos and then settled at the University of Chicago. He received the Nobel Prize for his demonstrations of the existence of new radioactive elements produced by neutron irradiation and for his work with slow neutrons.

Fermi, Laura. 1954. *Atoms in the Family: My Life with Enrico Fermi*. Chicago: University of Chicago Press. 267 p.

A warm, personal account of Fermi's life, this covers Fermi's early times in Italy as an energetic young man. The World War II years, Los Alamos, and especially the building of the first atomic pile are described. The science is described in language for the general reader. The politics and secrecy surrounding nuclear weapons are also discussed.

Latil, Pierre de. 1966. *Enrico Fermi: The Man and His Theories*. New York: Paul S. Eriksson, Inc. Translated by Len Ortzen. 178 p.

Both the personal and scientific sides of Fermi are covered, but the science is emphasized. There is an interesting discussion of pre-quantum physics, which sets the context for all the exciting discoveries of early 20th century physics and the events leading up to the development of the atom bomb. Fermi's role in all of this is well covered. The more or less uneventful years after World War II until his death, dedicated to teaching and thinking, are only briefly described.

Segrè, Emilio. 1970. *Enrico Fermi: Physicist*. Chicago: University of Chicago Press. 275 p.

Along with the conventional coverage of Fermi's life and his contributions to physics is an interesting discussion of how he worked and thought, and his approach to physics and solving problems. Much detail on how the first atomic pile was built is included. Fermi spent the last few years of his life as a public figure, as President of the American Physical Society and as one who testified at the Oppenheimer Security Hearings.

FEYNMAN, RICHARD PHILLIPS
Born May 11, 1918 (New York City)
Died Feb 15, 1988 (Pasadena)
Nobel Prize in Physics, 1965

An extraordinary person, with an enormous curiosity about absolutely everything, Feynman spent his early years in the Eastern United States. He was in Los Alamos during World War II and at Cornell University until 1950. He then moved to the California Institute of Technology, where he spent the rest of his career. He is perhaps best known for his course in introductory physics, *The Feynman Lectures in Physics* and for the development of quantum electrodynamics (for which he won the Nobel Prize).

Feynman, Richard P. 1985. *Surely You're Joking Mr. Feynman! Adventures of a Curious Character*. New York: W.W. Norton & Co. 350 p.

These autobiographical stories (as told to his long time friend and colleague Ralph Leighton) paint a picture of an amazing person hurling through life, determined to taste it all. Feynman had a wonderful sense of humor and was an inveterate mischief maker. Particularly amusing are the stories of himself as a safe cracker at Oak Ridge and Los Alamos during the days of the Manhattan Project. Because the science can't be truly separated from the scientist, many of the tales describe his approach to and discoveries in science.

Feynman Richard P. 1988. *What Do You Care What Other People Think? Further Adventures of a Curious Character*. New York: W.W. Norton & Co. 255 p.

As the title implies Feynman tells even more stories about his zest for life and desire to know (once again, as told to Ralph Leighton). The second half of the book details his work with the Presidential Commission investigating the Challenger disaster, including his dramatic demonstration at the hearings of what happens to the O Rings when immersed in ice water.

Leighton, Ralph. 1991. *TUVA OR BUST! Richard Feynman's Last Journey*. New York: W.W. Norton & Co.

Tannu Tuva, a land remembered only from Feynman's childhood hobby of stamp collecting, became the object of a determined quest for knowledge about that small part of the Soviet Union. The author and Feynman persisted through many adventures with the Soviet bureaucracy and other stumbling blocks, but Feynman died before he reached Tuva. Leighton and his wife did eventually travel to this promised land, and the final part of this book tells this tale.

"Special Issue: Richard Feynman." 1989. *Physics Today*. v42, n2, pp. 24-88.

Contributions from nine of Feynman's friends and colleagues present varied views of this remarkable physicist. The topics vary and include Feynman as a young man, the development of quantum electrodynamics, his style as a teacher, and his later work on the Connection Machine at the Massachusetts Institute of Technology. Also included are photographs of a selection of Feynman's artwork and the blackboards as he last wrote on them.

HEISENBERG, WERNER KARL
Born Dec 05, 1901 (Wurzburg, Ger)
Died Feb 01, 1976 (Munich)
Nobel Prize in Physics, 1932

Heisenberg was born in Germany and spent most of his life and career there. Another physicist with numerous interests other than physics, he studied philosophy and was much involved (in his later years) in the political side of science. He won the Nobel Prize for his fundamental work in the creation of quantum mechanics, but is probably best known for what has become termed the 'Heisenberg Uncertainty Principle' (that it is in principle impossible to determine both the position and momentum of a particle).

Cassidy, David Charles. 1976. *Werner Heisenberg and the Crisis in Quantum Theory, 1920-1925*. PhD Dissertation. Purdue University. 500 p.

Although primarily covering the beginning of Heisenberg's graduate studies up through his formulation of matrix multiplication (and therefore mostly quite technical), this does place Heisenberg's work within the history and sociology of his times. His connections with other physicists are also detailed.

Heisenberg, Elisabeth. 1984. *Inner Exile: Recollections of a Life with Werner Heisenberg*. Boston: Birkhauser. Translated by S. Cappellari and C. Morris. 170 p.

Heisenberg was one of the best known physicists to remain in Germany during World War II, and thus has been the recipient of much negativity. The principle aim of this book (written by his wife) is to try to explain and clarify Heisenberg's decision to stay in Germany. Details of his early years are given, as are his thoughts and relationships with others (both family and physicists). Lots of photographs add to the book.

Hermann, Armin. 1976. *Werner Heisenberg, 1901-1976*. Bonn-Bad Godesberg: Internationes. 145 p.

With many photographs and quotations of Heisenberg's own words, this book portrays a physicist with a wide range of interests. Particularly interesting are the discussion of the 'race' for the atomic bomb from a German's point of view and the coverage of the years of World War II.

JOLIOT-CURIE, IRÈNE
Born Sep 12, 1897 (Paris)
Died Mar 17, 1956 (Paris)
Nobel Prize in Chemistry, 1935

Few women scientists have had the good fortune to be raised in such a supportive intellectual atmosphere. Most of her life was spent in Paris, where, after a different sort of education, she received her doctorate while working at the Radium Institute (directed by her mother, Marie Curie). She and her husband Frédéric Joliot worked as a team much of their lives. They won the Nobel Prize for their discovery of artificially induced radioactivity. From 1946 to 1950 Joliot-Curie was one of the directors of the French Atomic Energy Commission, and spent her later years involved in other administrative and political activities.

Goldsmith, Maurice. 1976. *Frédéric Joliot-Curie: A Biography*. London: Lawrence and Wishart. 259 p.

A well rounded account of the life of Frédéric Joliot-Curie, this includes many details on his political actions in later years. There is a fair amount of coverage of Irène Joliot-Curie as well, including her life before the two met at the Radium Institute in Paris. Their contributions to science are described in detail, in language for the general reader.

McKown, Robin. 1961. *She Lived for Science: Irène Joliot-Curie*. New York: Julian Messner, Inc. 192 p.

The life and professional career of Joliot-Curie are described in this personal account, with emphasis on her being the daughter of the famous Marie and Pierre Curie. Her childhood and early education, her work alongside her mother in World War I with the new X-Ray machines, her work at the Radium Institute, and her marriage and partnership with Frédéric Joliot are well covered. This is written for young adults, and the science is perhaps too simplified, but the adult reader will still benefit.

Opfell, Olga S. 1986. "Triumph and Rebuff: Irène Joliot-Curie." In: *The Lady Laureates: Women Who Have Won the Nobel Prize*. Metuchen, NJ: Scarecrow Press, Inc. pp. 195-212.

In a brief biographical sketch, all facets of Joliot-Curie's personality and all the periods of her life are covered. Good coverage of her scientific work in radioactivity is given.

LANDAU, LEV DAVIDOVICH
Born Jan 22, 1908 (Baku, Rus)
Died Apr 01, 1968 (Moscow)
Nobel Prize in Physics, 1962

A genius with a love for classifying and labelling all manner of things (including physicists; Einstein was the highest and had no match), Landau was born, raised, and educated in Russia. Beginning in 1929, he spent a fruitful two years visiting European centers of physics, learning from and working with many of the greatest physicists of those days. Landau was never a specialist and worked in many areas of physics. He won the Nobel Prize for his work with superfluid helium, and was well known for his three volume work (with E.M. Lifshits) on quantum mechanics. Landau was nearly killed in an automobile accident, in 1962, and spent the last years of his life living quietly, unable to continue his scientific work.

Dorozynski, Alexander. 1966. *The Man They Wouldn't Let Die*. London: Secker & Warburg. 207 p.

Landau's life is described in glowing terms, covering his early days through his years as a gifted teacher. There are interesting details about practicing science in Stalin's reign; absurdities and tragedies abounded (Landau himself spent over a year in jail as a political prisoner). There is much coverage of the incredible outpouring of love and giving of everything possible after the accident, and some details of his remaining years.

Khalatnikov, I.M. 1989. *Landau: The Physicist and the Man. Recollections of L.D. Landau*. Oxford: Pergamon Press. Translated by J.B. Sykes. 323 p.

Numerous short reminiscences of a great man constitute this book. Although mostly biographical, there is also coverage of Landau's physics and discussion of how he thought and worked.

Livanova, Anna. 1980. *Landau: A Great Physicist and Teacher*. Oxford: Pergamon Press. Translated by J.B. Sykes. 217 p.

With some discussion of his early life, this concentrates on Landau's life as a physicist and in particular, his gift for being a teacher. Landau studied the whole of physics and published in many subfields. The second half of the book is an indepth description of Landau's contributions to the theory of superfluidity and liquid helium (in language for the nonspecialist).

LAWRENCE, ERNEST ORLANDO
Born Aug 08, 1901 (Canton, SD)
Died Aug 27, 1958 (Palo Alto)
Nobel Prize in Physics, 1939

American born and raised, Lawrence first taught at Yale University, and left that august institution for the relatively unknown University of California at Berkeley in 1928. His tenure there saw its rise to a renowned center of physics. He is well known for his invention and development of the cyclotron (for which he won the Nobel Prize) and for his work on the Manhattan Project during World War II.

Childs, Herbert. 1968. *An American Genius: The Life of Ernest Orlando Lawrence*. New York: E.P. Dutton & Co., Inc. 576 p.

Scholarly and detailed (yet written for the nonphysicist) this covers Lawrence's life, from his early years through his untimely death. His cyclotron is described indepth (with photographs and diagrams) and the constant struggles for money to build ever bigger ones are documented. His work through the years of World War II and afterwards is included. Many anecdotes and photographs provide a portrait of a gifted experimental physicist, always striving for more.

Davis, Nuel Pharr. 1969. *Lawrence and Oppenheimer*. London: Jonathon Cape. 385 p.

This somewhat controversial dual biography draws comparisons and contrasts between Lawrence, a remarkable experimental physicist, and J.R. Oppenheimer, the theoretician (the controversy stems from the author's portrayal of Lawrence in particular). The two worked closely together both at the University of California at Berkeley and at Los Alamos during the Manhattan Project. Their interactions with many other physicists are also detailed. Coverage continues through the Oppenheimer Security Hearings and the awarding of the Fermi Award to Lawrence in 1957.

Heilbron, J.L. and Robert W. Seidel. 1989. *Lawrence and His Laboratory: A History of the Lawrence Berkeley Laboratory. Volume 1*. Berkeley: University of California Press. 586 p.

The early years of the Lawrence Berkeley Laboratory through 1941 (Pearl Harbor) are detailed, and well illustrated with photographs and diagrams. While Lawrence is not the true focus of this book, he was the driving force behind the Laboratory, and biographical information is provided. Volume 2 will cover World War II through the early years of the Cold War and the Korean War. Volume 3 will cover the evolution of the modern national laboratory.

MAYER, MARIA GOEPPERT
Born Jun 28, 1906 (Kattowitz, Pol)
Died Feb 20, 1972 (La Jolla)
Nobel Prize in Physics, 1963

The second woman to ever win the Nobel Prize in Physics, Mayer was from a long line of university professors. She grew up in the German university town of Göttingen, one of the centers of the new quantum physics. Earning her PhD and getting married in 1930, she moved (with her American husband) to the United States, where she held a succession of mostly unpaid positions at various universities. She worked on some aspects of the Manhattan Project, and in Chicago, both at the University and at Argonne National Laboratory, she developed her model of the nuclear shell structure, for which she won the Nobel Prize (shared with J.H.D. Jensen, who independently developed the same model). It wasn't until she went to the University of California at San Diego in 1960 that she finally attained a fully paid position as a professor.

Dash, Joan. 1973. *A Life of One's Own: Three Gifted Women and the Men They Married*. New York: Harper & Row, Pub. pp. 229-346.

The life of Mayer, from her early education through her PhD and beyond is covered. Her scientific contributions, in particular the nuclear shell model, are described. Included is much commentary on the problems encountered by women in the field of physics and some discussion of why so few women enter science.

Johnson, Karen Elise. 1986. *Maria Goeppert Mayer and the Development of the Nuclear Shell Model*. PhD Dissertation. University of Minnesota. 350 p.

Tracing both the life of Mayer and the development of the nuclear shell model by Mayer and others, this provides coverage through 1955 (when the model was introduced). Her scientific achievements, from her PhD on are described, including the coauthored textbook *Statistical Mechanics* and all the work leading up to the nuclear shell model. The portrayal of Mayer as a woman physicist is low key, never overly dramatic.

Opfell, Olga S. 1986. "Madonna of the Onion: Maria Goeppert Mayer." In: *The Lady Laureates: Women Who Have Won the Nobel Prize*. Metuchen, NJ: Scarecrow Press, Inc. pp. 224-238.

A brief biographical sketch, this covers the major events of the life of Mayer from her early years and marriage to an American, and the stints at various universities and Argonne National Laboratory. Her scientific work and how she worked with other physicists are also described.

MEITNER, LISE
Born Nov 07, 1878 (Vienna)
Died Oct 27, 1968 (Cambridge)

An Austrian by birth, Meitner was among the first women to enter the University of Vienna (in 1901) and the first to receive a doctorate in physics there. After moving to Berlin to further her studies, she began to work on the new radioactivity with Otto Hahn, a working partnership and friendship that was to last a lifetime. Meitner and Hahn discovered a new element, proactinium, in 1918. She is probably best known for the explanation (with Otto Frisch) of some curious experimental results; they realized that the nucleus had actually split in two (or fissioned). That she did not at least share the Nobel Prize for this discovery is not easily explained. It was this realization of fission that led to the atomic bomb (and more peaceful uses of atomic energy).

Crawford, Deborah. 1969. *Lise Meitner: Atomic Pioneer*. New York: Crown Pub. 190 p.

A well written account of Meitner's life and work, this includes good coverage of physics and Meitner's scientific contributions. It portrays a woman who struggled to be recognized for her talents, yet still enjoyed her life and friendships with other physicists. Also discussed are the moral issues of scientific discoveries leading towards instruments of death. While written for young adults, this biography will also be of interest to adults.

Hermann, Armin. 1979. *The New Physics: The Route Into the Atomic Age. In Memory of Albert Einstein, Max Von Laue, Otto Hahn, Lise Meitner*. Bonn-Bad Godesberg: Internationes. 175 p.

An interesting book, in a rather terse style, this covers four physicists who were instrumental in ushering in the atomic era. Details of Meitner's life and work are provided. There are many good quality photographs and reproductions of many famous seminal papers. The last chapter is a chronology of events from 1878 (Meitner's birth) through 1968 (Meitner's death). Four separate topics are juxtaposed: the milestones of each scientist's career, general science and technology, politics and society, and culture.

Rife, Patricia Elizabeth. 1983. *Lise Meitner: The Life and Times of a Jewish Woman Physicist*. PhD Dissertation. Union Graduate School. 505 p.

A narrative of Meitner's life, this focuses on the influences of her being Jewish and a woman. Her life is set into the context of 20th century social history and the history of science. The science is covered adequately for the nonphysicist, and society and culture are covered in detail.

MICHELSON, ALBERT ABRAHAM
Born Dec 19, 1852 (Strelno, Prussia)
Died May 09, 1931 (Pasadena)
Nobel Prize in Physics, 1907

Michelson's parents emigrated to the United States when he was a young boy. They joined the gold rush stampede, but as merchants and suppliers to the prospectors. Gaining higher education through an appointment to the U.S. Naval Academy, Michelson embarked on his career as an experimental physicist. He is best known for a succession of experiments that measured the speed of light (in collaboration with Morley, hence the Michelson-Morley Experiment) and for inventing the interferometer. He won the Nobel Prize, the first American to win one in the sciences, for the investigations he carried out with the aid of his precision optical instruments.

Goldberg, Stanley and Roger H. Stuewer (Eds.). 1988. *The Michelson Era in American Science 1870-1930*. New York: American Institute of Physics. (*AIP Conference Proceedings* 179). 300 p.

This collection is dedicated to the 100th anniversary of the Michelson-Morley Experiment. Along with measuring the speed of light, it gave a null result on the existence of ether, long thought to be a necessary medium for the propagation of light. While there is some debate on the true importance of the experiment, it did provide the basis of Einstein's theory of relativity. There is coverage of Michelson's influence on the schools where he taught and on science in general in America, as well as interesting commentary on the roles of experiment and theory.

Hughes, Thomas Parke. 1976. *Science and the Instrument-Maker: Michelson, Sperry, and the Speed of Light*. Washington: Smithsonian Institution Press. (*Smithsonian Studies in History and Technology*, no. 37). 18 p.

Providing brief biographical sketches of Michelson and Sperry (a gifted inventor), this focuses on the interaction of the two men in developing the instrument needed for ever more precise measurement of the speed of light beginning in 1924 (after the more famous Michelson-Morley Experiment). The experiment itself, including troublesome differences in data reporting, is covered in detail.

Livingston, Dorothy Michelson. 1973. *The Master of Light: A Biography of Albert A. Michelson.* New York: Charles Scribner's Sons. 376 p.

Written by the daughter of Michelson and his second wife, this a relatively balanced account of the personal life and scientific contributions of Michelson. There are many diagrams and photographs. Also included is a rather interesting discussion of the fact that while Michelson himself had essentially disproved the theory of ether, he could never quite give up the idea and accept wholeheartedly the new theories of relativity and quantum physics.

OPPENHEIMER, J. ROBERT
Born Apr 22, 1904 (New York City)
Died Feb 18, 1967 (Princeton)

A fascinating person, Oppenheimer was born and raised in an upper class family in New York City. He didn't begin in depth studies in physics until his early twenties when he went to Europe to study. He earned his PhD in 1927, and in 1929 left for California, where he accepted part-time positions at both the University of California at Berkeley and the California Institute of Technology. He continued to develop important aspects of quantum physics and was well regarded as a teacher. He was involved in early discussions of the feasibility of developing an atomic bomb and in early 1942 was named coordinator of fast fission research. In November 1942 he was named Director of the Los Alamos Laboratory, a task for which he exhibited an unexpected genius. After World War II he became Director of the Advanced Study Institute in Princeton, where he worked until his death. In 1953 he was informed his security clearance was being revoked, and thus began the Oppenheimer Security

Hearings, one of the most infamous trials of the McCarthy anti-Communism hysteria.

Goodchild, Peter. 1983. *Oppenheimer: The Father of the Atomic Bomb*. London: British Broadcasting Corp. 303 p. (originally published as *Oppenheimer: Shatterer of Worlds* in 1980).

Portraying a person with many interests outside physics (including learning Sanskrit, from which came one of his most famous quotes, after witnessing the first test of the atomic bomb: "I am become death, destroyer of worlds.") this covers the whole of Oppenheimer's life, with great attention to background details. The Manhattan Project and events leading up to it, and the Security Hearings are covered indepth. Also included are more personal aspects, such as a psychological interpretation of his boyhood and relationships with women. Numerous photographs add to the book.

Kunetka, James W. 1982. *Oppenheimer: The Years of Risk*. Englewood Cliffs, NJ: Prentice-Hall, Inc. 292 p.

Primarily covering the years 1942-1954, the years of government service for Oppenheimer, this traces his direct influence on the development of the first and subsequent atomic weapons. The Security Hearings, with extensive quotes, are covered in detail. A brief sketch of Oppenheimer's early years and photographs round out the picture.

Larsen, Rebecca. 1988. *Oppenheimer and the Atomic Bomb*. New York: Franklin Watts. 192 p.

While there is some coverage of Oppenheimer's earlier years, this concentrates on the activities during World War II at Los Alamos and afterwards. The Security Hearings are described in detail, as are his thoughts on the role and responsibilities of scientists. While written for young adults, this can also be enjoyed by the adult reader.

Michelmore, Peter. 1969. *The Swift Years: The Robert Oppenheimer Story*. New York: Dodd Mead & Co. 273 p.

In a sympathetic approach, this is the story of the life of Oppenheimer, with emphasis on how he was fashioned by his times and how his times were fashioned by him. Oppenheimer is described as being very hard to characterize, he left few private papers and people had quite different opinions about him. While there is some coverage of his early years and the years at Berkeley with Lawrence, the primary focus is on the years of World War II and Oppenheimer's statesmanlike role afterwards.

RABI, ISIDOR ISAAC
Born Jul 29, 1898 (Rymanow, Aus)
Died Jan 11, 1988 (New York City)
Nobel Prize in Physics, 1944

Born in Austria, Rabi grew up and was educated in America. He went to Europe after his PhD to study the new quantum physics. He was a professor at Columbia University from 1929 until his retirement in 1967. During World War II he was one of the scientists working on the development of microwave radar and was Associate Director of the Radiation Laboratory in Cambridge, MA. He was later active in promoting peaceful uses for atomic energy and very involved as a spokesperson for science. He won the Nobel Prize for his atomic and molecular beam work and for determining the magnetic moments of fundamental particles.

Bernstein, Jeremy. 1975."Profiles: I.I. Rabi: Physicist, parts 1 and 2." *New Yorker*. vol. 51, no. 2, October 20, pp. 47-110 and no. 3, October 27, pp. 47-101.

In an informal style, Rabi's scientific achievements and involvement in politics are covered in this two part article. His youth and education, his work at the Radiation Laboratory, the research that resulted in the Nobel Prize, and especially his identification with Columbia University are discussed.

Rigden, John S. 1987. *Rabi: Scientist and Citizen*. New York: Basic Books, Inc. 302 p.

A well written and balanced account of an admirable person, this emphasizes the views of Rabi that scientists have a responsibility to society (it was at his suggestion that President Eisenhower create the office of Science Advisor to the President). His early years, in which religion, poverty, and ethnic isolation all exerted great influence, are described. The other major periods of his life — his work on the Manhattan Project and at the Radiation Laboratory, professor at Columbia, testifying at the Oppenheimer Security Hearings, and involvement in politics are covered as well.

YUKAWA, HIDEKI
Born Jan 23, 1907 (Tokyo)
Died Sep 08, 1981 (Kyoto)
Nobel Prize in Physics, 1949

The first Japanese to win the Nobel Prize in Physics, Yukawa was born, raised, and educated in Japan. He began teaching at Kyoto Imperial University, taught at Osaka Imperial University, and then moved back to Kyoto University in 1939, where he worked until he retired in 1970. He spent a number of years in the late 1940's and 1950's as a visiting professor at institutions in America. In his later years he was very involved in protesting nuclear weapons and campaigning for nuclear disarmament. He is best known for his prediction of the meson on the basis of theoretical work in nuclear forces, for which he won the Nobel Prize.

Brown, Laurie M. 1986."Hideki Yukawa and the Meson Theory."
Physics Today. v39, n12, pp. 55-62.

The development of the prediction (by Yukawa) and discovery of
the meson is traced, including the confusion in physics in the early
1930's over the composition of the nucleus (proving the existence
of the meson helped to clear the situation). The article is quite technical, but the portrayal of how Yukawa worked, approached problems, and interacted with colleagues is interesting.

Yukawa, Hideki. 1982. *Tabibito = Traveler*. Singapore: World
Scientific Pub. Translated by L. Brown and R. Yoshida. 218 p.

Autobiographical reminiscences provide an endearing portrait of
a physicist's journey through life. Focusing mostly on his early
years through 1934 (when he predicted the meson), this covers his
family life, school days, and professional life as a physicist. A
chronology and several photographs are included.

SOURCES

Boorse, Henry A., Lloyd Motz, and Jefferson Weaver. 1989. *The Atomic Scien-
    tists: A Biographical History*. New York: Wiley.
Dictionary of Scientific Biography. 1970-1980. New York: Scribner.
McGraw-Hill Modern Scientists and Engineers. 1980. New York: McGraw-Hill.
Schlessinger, Bernard S. and June H. Schlessinger. (Eds.) 1986. *The Who's Who
    of Nobel Prize Winners*. Phoenix: Oryx Press.
Wasson, Tyler (Ed.) 1987. *Nobel Prize Winners: An H.W. Wilson Biographical
    Dictionary*. New York: H.W. Wilson.

Lives of Chemists

Gayle Baker
Marie Garrett

In John Elmsley's review of the book *Aleksandr Porfir'evich Borodin*, he states, "Chemistry has a habit of breaking out in the most unusual places, stirring a group of young people to action and producing world-changing discoveries."[1] The chemists whose names are gathered on these pages form a diverse group. They are people from different places and different times who reflect a wide variety of research interests. They are united, however, in the two respects that Elmsley mentions: action and discovery. These chemists established research goals and tenaciously pursued them, whether in academia or in industry. They shared their knowledge and the results of their research through teaching, speaking, and writing. Active not only in the laboratory but also in the world, some of them are better known for their contributions in fields unrelated to science. Many of their research projects were geared toward meeting practical needs—developing new methods and products, serving their countries during times of war, and searching for cures to life-threatening diseases. Many of their scientific quests resulted in discoveries that changed the world. For their contributions, these

Gayle Baker is Coordinator of Reference Services for Science and Technology at Hodges Library, University of Tennessee, 1015 Volunteer Blvd., Knoxville, TN 37996-1000. She holds a Bachelor's degree in Mathematics from Bowling Green State University and a Master's degree in Computing and Information Science from Ohio State University. Her Master's in Library Science is from the University of Alabama.

Marie Garrett is currently Reference Librarian at Hodges Library, University of Tennessee, 1015 Volunteer Blvd., Knoxville, TN 37996-1000. She holds a Bachelor's degree in English from Milligan College, Tennessee; Master's degree in English from the University of Tennessee and the MLS from George Peabody College.

chemists received numerous honors and awards, some of them even earning Nobel Prizes. They deserve recognition not only for their discoveries but also for their daily commitment to science and to humanity. Perhaps this dedication is one of the primary characteristics that has inspired biographers to explore the lives of these noted chemists in depth. No doubt, a desire to recognize daily commitment has played a part in Editor Jeffrey L. Seeman's and the American Chemical Society's publication of the autobiographies of twentieth-century chemists.[2]

Aleksandr Porfir'evich Borodin: A Chemist's Biography. By N. A. Figurovskii and Yu I Solov'ev. Trans. from the Russian by Charlene Steinburg and George B. Kauffman. Berlin: Springer-Verlag; 1988. 171 p. Photographs. Indexes.

The Russian name Aleksandr Porfir'evich Borodin (1833-1887) is more often associated with music than with science. But the man who composed the melody that accompanies the lyrics to "Stranger in Paradise" also developed the aldol condensation reaction and devised a method for testing urea that was used in the medical profession for many years. Music was more often something that Borodin hummed while involved in laboratory research than something that he composed. *Prince Igor*, an opera that he worked on over a span of eighteen years, remained unfinished when he died at age 53. This chemist and teacher described himself as being "completely happy" (p. 64) while involved in research, but he readily interrupted his own work "without impatience" (p. 50) to advise or assist students and colleagues. As a faculty member at the Medical-Surgical Academy in St. Petersburg, he also took time from other endeavors to fight for the inclusion of women's courses in the curriculum. Figurovskii and Solov'ev do not neglect Borodin's contributions to music, but they emphasize the "enormous significance" (p. 71) of his work in organic chemistry. Their work relies heavily on letters he wrote to his wife. The brief, readable chapters are supplemented by lists of Borodin's chemical and musical works and the texts of some of his reports, correspondence, speeches, and conference papers. (MG)

Chaim Weizmann: A Biography. By Norman Rose. New York: Viking; 1986. 520 p. Photographs. Index.

Born in Motol, Russia, Chaim Weizmann (1874-1952) lived to see his name become known throughout the world. As a young boy, he developed a desire to see the Jewish people return to their Palestinian homeland. As an adult, he became a strong Zionist leader and played a significant role in the events which led to that return. In 1949, Weizmann became Israel's first president. But along the path to that distinction, he also pursued a career in science. Inspired by his own chemistry teacher, he taught chemistry at England's Manchester College for ten years. Weizmann found "great mental and physical relaxation" in conducting research, including work on "high-octane aviation fuel and synthetic rubber . . . medicinal drugs . . . substitute foodstuffs . . . [and] acetone production" (p. 375). Some of this research was especially important during wartime. The Weizmann Institute of Science that he founded "personified his conception of the role Zionism could play as a force in Israel and in the modern world: erudition and vision combined with practicality and purpose, learning harnessed to the needs of society" (p. 3). Norman Rose's own introduction to this leader's importance took place at a memorial service following Weizmann's death. His subsequent interest in the man has resulted in a thoroughly researched and well documented biography that supplements Weizmann's own autobiography, *Trial and Error*. (MG)

From Cologne to Chapel Hill. By Ernest L. Eliel. Washington: American Chemical Society; 1990. 138 p. Photographs. Index.

"During the summer at OSU in 1952, I had been thinking about the conformational analysis of mobile systems, such as nucleophilic displacements on cyclohexyl bromide" (p. 30). So begins a chapter in the autobiography of Ernest Ludwig

Eliel (1921-). His reflections on chemical matters have led to the publication of numerous articles. One of his textbooks, *Stereochemistry of Carbon Compounds*, sold over 41,000 copies in twelve printings and "was translated into Japanese, German, Czech, and Russian" (p. 45). Born in Cologne, Germany, Eliel studied in Edinburgh, Scotland, then escaped Nazi internment camps to study in Havana, Cuba. After coming to the United States, he received a Ph.D. in chemistry from the University of Illinois. Most of his teaching career took place at the Universities of Notre Dame and North Carolina. He also did some research in industry and served as president of the American Chemical Society. Eliel readily gives credit to his colleagues and to "graduate students, postdoctoral associates, and undergraduate researchers" (p. 112) who have contributed to his own work. Of the many honors he has received, he selected three that are of special significance to him: the University of North Carolina's Award in Science for 1986, a 65th birthday party-symposium for him and his colleague Bob Parr and an honorary doctorate conferred by Notre Dame in 1990. His autobiography provides a glimpse into the workings of a chemist's mind and a sense of his perspective about what is important in his work. Eliel's memoirs form the second volume in the American Chemical Society's *Profiles, Pathways, and Dreams* series. (MG)

Grand Obsession: Madame Curie and Her World. By Rosalynd Pflaum. New York: Doubleday; 1989. 496 p. Photographs. Index.

Marie Sklodowski Curie (1867-1934) "believed so strongly that there was only one source of progress—science—" (p. 232) that she devoted her life to research. Making "scientific progress to benefit all humanity" (p. 295) became a "grand obsession" not only for this Polish chemist but also for others of her family members. Nobel Prizes were awarded to the Curie family three times: in 1903 to Marie and her husband Pierre, a French physicist, for their discovery of radium and polonium; in 1911 to Marie for her isolation of pure radium; and in 1935 to the Curie's daughter and son-in-law for their synthesis of artificial radiation. That first discovery and "Marie's daring hypothesis that radiation was an atomic property, are two important milestones in the opening of the atomic age and nuclear physics" (p. 92-93). Humanity has benefited from the use of radium for the treatment of cancer. Rosalynd Pflaum's biography of this remarkable family focuses on the Curie's active involvement in their world and the personal sacrifices they willingly made in pursuit of scientific progress. During her research, she interviewed colleagues, family members, and friends of the Curie's. Her selected but substantial bibliography includes these interviews and other unpublished material as well as an impressive list of print sources. (MG)

Lavoisier and the Chemistry of Life: An Exploration of Scientific Creativity. By Frederic Lawrence Holmes. Madison, WI: University of Wisconsin Press, 1985. 565 p. Illustrations. Photographs. Index.

Frederic L. Holmes provides a most scholarly look at the research of Antoine Lavoisier during the years 1773 through 1792. During this time, Lavoisier studied respiration, fermentation, and the chemistry of plants and animals. Among the highlights of his work during this time are the research with the French mathematician Laplace in calorific theory and animal heat, the publication of *Methode de nomenclature chimique* with Morveau, Berthollet and Fourcroy, the research on respiration with Sequin, and the publication of his textbook, *Elementary Treatise of Chemistry.* The author states that his purpose is to focus "on the intrinsic character of Lavoisier's scientific work rather than on the social context within which he pursued it" (p. 501). Holmes gained access to the archives of the Academie des Sciences where he was able to use the laboratory notebooks, notes, memoranda, and manuscripts of Lavoisier. Detailed descriptions of many of these documents are provided. The most interesting aspect of this book is the use of successive drafts of manuscripts in order to analyze the evolution of Lavoisier's ideas. Detailed notes are supplied for each chapter. (GB)

Linus Pauling: A Man and his Science. By Anthony Serafini. New York: Paragon House; 1989. 310 p. Photographs. Index.

This biography of Linus Pauling, winner of two Nobel Prizes, covers the life of one of the most noted, most colorful, and sometimes, controversial scientists living today. Pauling received his Ph.D. in chemistry at Caltech, joined the faculty, and later became chairman of the Department of Chemistry and Chemical Engineering. His research covered many areas: chemical bonds, the race for the structure of DNA, sickle-cell anemia, the relationship between nutrition and mental health, Vitamin C and the common cold. Pauling wrote a landmark book, *The Nature of the Chemical Bond*. His work in this area earned him the Nobel Prize in chemistry in 1954. Serafini presents a negative side of Pauling and his work. While Pauling has been regarded as a brilliant theorist, he has been accused by his detractors as being deficient in producing data in support of his theories. When performing research in an area outside of chemistry, such as medicine, Pauling has been criticized for his lack of knowledge and training in the field. He is passionate, not only about his scientific theories, but also about his beliefs in nuclear proliferation. These activities outside of chemistry caused him many problems, both legal and political. Nevertheless, he was awarded a second Nobel Prize in 1964 for peace, and, in 1975, the National Medal of Science. The author used transcripts of interviews, made by others, of Pauling's colleagues, as well as interviews of his own. The majority of citations to printed material are from newspapers. (GB)

The Right Place at the Right Time. By John D. Roberts. Washington: American Chemical Society, 1990. 299 p. Photographs. Index.

In his autobiography, John D. Roberts writes that "Being at the right place at the right time, usually with my hands at least half open to latch on to the goodies that sail by on fortune's wind, has led to a fun-filled, variety-filled career in education and research" (p. 258). Among the early opportunities that this California-born chemist seized were learning the techniques of glass-blowing, serving as an undergraduate teaching assistant, and starting research as an undergraduate at UCLA in the 1940s. World War II interrupted the start of his graduate studies at Penn State and brought him back to UCLA to become involved in war-related research on the production of oxygen. Roberts was able to complete research on butenyl Grignard for his Ph.D. at UCLA without having to take any graduate courses in chemistry. A National Research Council fellowship brought him to Harvard on a post-doc, and he eventually joined the chemistry faculty at MIT, although continuing to maintain his ties with the Harvard faculty. During this time he became known for his work in molecular orbital theory. A Guggenheim fellowship brought him to Caltech at an opportune time. He became the primary candidate to fill an opening in their chemistry department. He joined the Caltech faculty in 1953, and later on, began his work in nuclear magnetic resonance (NMR) spectroscopy for which he is best known. The author provides insight into the people and papers which influenced Roberts in his work, and includes interesting vignettes which reveal the personalities of his colleagues and students. A large portion of the book is devoted to discussion of Roberts' research. This volume is the first of the American Chemical Society series *Profiles, Pathways, and Dreams*. (GB)

Robert Robinson: Chemist Extraordinary. By Trevor I Williams. Oxford: Clarendon Press; 1990. 201 p. Photographs. Indexes.

Sir Robert Robinson (1886-1975) maintained a diversity of interests throughout his eighty-nine years. He won the Nobel Prize for chemistry in 1947 and presided over the Royal Society in England from 1945 to 1950, but he also served as president of the British Chess Federation and was elected as a Knight of Mark Twain. While teaching chemistry in Australia, Scotland, and England, he actively pursued his interests in mountaineering, photography, gardening, chess and piano playing. His research took place not only in academic settings but also in industries such as the British Dyestuff Corporation and Shell Research Company. His research produced hundreds of articles and advancements with such diverse products as penicillin, petroleum, flower pigments, hormones, alkaloids, steroids, synthetic antimalarials and antiseptics, and chemical warfare explosives. The Nobel Prize was awarded for his work with morphine and strychnine. Trevor Williams had been personally acquainted with Robinson and enthusiastically accepted an invitation from Marion Robinson Way to write her father's official biography. Williams had access to the first volume of Robinson's autobiography, *Memoirs of a Minor Prophet*, and other printed sources. But unpublished letters and the personal recollections of many of Robinson's colleagues and friends have helped him to create a picture of Robinson's personal as well as his scientific life. (MG)

Roger Adams: Scientist and Statesman. By D. Stanley Tarbell and Ann Tracy Tarbell. Washington: American Chemical Society; 1981. 240 p. Photographs. Index.

Roger Adams, a descendant of the famous Adams family of Presidents, pursued his graduate and undergraduate studies at Harvard. In 1912 he received a fellowship for study at the University of Berlin, where he "learned what it took to do first rate research" (p. 30). In 1916 Adams joined the faculty of the University of Illinois where he became involved with the famous Illinois "preps" labs, where large quantities of key compounds were prepared that were previously supplied by the Germans. Adams worked on perfecting techniques for synthesizing these compounds and this proved to lay the groundwork for the series *Organic Syntheses*. After World War I, Adams worked on building his own research program, and in 1922 he discovered "Adams platinum oxide," a catalyst for the reduction of multiple bonds. From 1927 to 1942, he served as head of the chemistry department at Illinois. His professional activities included serving as president of the American Chemical Society. Perhaps his greatest accomplishments were made during the 1940s with his work on the National Defense

Research Committee, where he was able to use both his chemical and administrative expertise. He was asked to visit postwar Germany as an advisor and to attempt to resuscitate German scientific publications. Adams also visited postwar Japan and "opened to the Japanese a new view of research, teaching, and the role of fundamental science in supporting technology" (p. 163). His postwar research focused on natural products. Among his awards were the Priestly Medal (ACS) and the National Medal of Science. The authors used the documents from various archives and interviewed and/or corresponded with many of Adams's colleagues. (GB)

The Story of Fritz Haber. By Morris Goran. Norman, OK: University of Oklahoma Press; 1967. 240 p. Photographs. Index.

In the preface, Morris Goran states that Fritz Haber "was recognized as the authority in the relations between science and industry, science and government, and science in support of war and peace" (p. vii). Haber studied at several German universities, but eventually pursued his doctoral studies, on indigo and associated substances, at the Charlottenburg *Technische Hochschule*, home of the largest engineering college in Germany. After some undistinguished work in industry, Haber received a faculty appointment to the Karlsrhue *Technische Hochschule* in the Department of Chemical and Fuel Technology. In 1898, he published *Outline of Technical Electrochemistry on a Theoretical Basis*, described as "an effective union of the technical and the theoretical" (p. 26). Haber's synthesis of ammonia brought him a controversial Nobel Prize in 1918, when he received criticism for his work on poisonous gas during World War I. Haber was diligent in his effort for postwar Germany in his work with the Emergency Society for German Science. In order to help the fatherland pay its war debts, he engaged in research into the extraction of gold from sea water that proved to be economically unfeasible. His work at the *Dahlen Kaiser Wilhelm Institute* in the area of physical chemistry and electrochemistry was successful due to his ability to bring together scientists from many different fields to work on a problem. Jewish by birth, Haber was forced to leave the institute by the Nazis and lived the remainder of his life in exile. In addition to the bibliography, the book includes a list of Haber's publications and lectures. (GB)

REFERENCES

1. Elmsley, John. "Kismet Chemist." *New Scientist* 121 (4 Feb. 1989): 68.

2. Seeman, Jeffrey L., ed. *Profiles, Pathways, and Dreams: Autobiographies of Eminent Chemists*. Washington: American Chemical Society; 1990-91. (A series designed to trace the recent history of organic chemistry through the lives of some of the people who created that history. The 22-volume series is scheduled for completion in 1991. Two of the volumes are reviewed in this article.)

Biography in the Geological Sciences

Flossie E. Wise

Geology as a recognized science is relatively young when compared to mathematics or chemistry. Much of its past has been influenced by myths or religious beliefs, and not until the late eighteenth century, did pieces of the puzzle that is earth science begin to fall into place. How the earth is structured, the forces at work within it, its surfaces and depths, millions of years of movements and upheavals, fossil records—these are all parts of the study of the geological sciences. Its study and development has been closely intertwined with the work of botanists, chemists, and physicists. Many early studies and discoveries were not made by geologists, but by others who happened to come across interesting quirks in the earth and delved further, hoping to find an explanation of mysteries that had confounded scientists for centuries. Earthquakes, floods, and volcanic activity have been attributed to many mysterious causes. For example, Aristotle declared that earthquakes were the result of underground winds, a view that was common for almost 2000 years. Volcanoes were thought to be the result of supernatural events or, depending on the culture, were the retribution of angry gods. Fossils were said to be unsuccessful attempts by nature to form plants and animals.

Geology came of age in Britain, and British theories and methods quickly spread to other parts of the world. By the end of the nineteenth century, we see a distinct trend toward the economic aspects of geology, first by the British and then by the Americans. In the

Flossie E. Wise is Reference Librarian in Hodges Library, University of Tennessee, 1015 Volunteer Blvd., Knoxville, TN 37996-1000. She holds a Bachelor's degree in Home Economics and a Master's in Librarianship from the University of Tennessee.

United States, the U. S. Geological Survey had been established and geologists Powell, King, and Hayden were conducting surveys in the West. Mines opened, dinosaur remains were unearthed, oil wells were tapped, and geology became a respected science. By the twentieth century, expansion in related areas of the earth sciences began to flourish. New tools and techniques were developed, and continental drift and plate tectonics became the "topics of the day."

The biographies chosen for this exercise represent a fraction of those available. Some are of major figures in the development of geology. Others may not be so important, but they are exciting to read.

Bones for Barnum Brown: Adventures of a Dinosaur Hunter. By Roland T. Bird. Fort Worth: Texas Christian University Press; 1985. 225 p. Illustrations. Notes. Index.

This attractive book reads like an adventure story. R. T. Bird (1899-1978) writes of his experiences as right-hand man to Barnum Brown (1873-1963) during the 1930's. Brown was known as a hunter of dinosaurs, and Bird was considered "one of the last of the old time fossil hunters" (p. vii). To provide background, the history of the study of dinosaurs is reviewed. Brown enters the picture around 1902 when he started working in eastern Montana on Late Cretaceous beds, collecting specimens of herbivorous dinosaurs. He was considered one of the "big three" among fossil hunters of the time. He did extensive work in Canada, but also travelled to Cuba, Mexico, India, Pakistan, and the Mediterranean. In 1927 he became Curator of Vertebrate Paleontology at the American Museum, where he stayed for sixty-six years. Bird and Brown first met around 1932, and their working relationship lasted until Bird's poor health brought it to a halt at the end of World War II. Bird's writing is vivid. "The gizzard stone of a dinosaur. . . . I said it out loud. I may have shouted it. A millions-of-years-long time-wave swept over me there in the desert, and I could see the great reptile picking up this stone, attracted by its bright color, carrying it perhaps his life long, dying here"

(p. 39). Not only is this an exciting tale to read, but the drawings of dinosaurs are wonderful. There is an extensive notes section, which is intended to lead the reader to additional sources about topics discussed by Bird. The annotations and the introduction were prepared by James O. Farlow, since Bird died before his book could be published.

Charles Lyell. By Sir Edward Bailey. Garden City, NY: Doubleday and Co.; 1963. 214 p. Photographs. Illustrations. Index.

"It is certain that to travel is of threefold importance to those who desire to originate just and comprehensive views concerning the structure of this globe" (p. 83). During his lifetime, Charles Lyell (1797-1875) traveled widely and made considerable geological observations, not only in his native Scotland, but also in Europe and America. "He steadfastly sought truth through deduction from observation" (p. 204). In this respect he was much like James Hutton, and Lyell's theories closely followed Hutton's in that he believed the earth had changed slowly and gradually. The author also details a close association with geologists Roderick Murchison and Adam Sedgwick, and in particular, looks at Lyell's friendship with Charles Darwin. Included are excerpts from letters between the two that provide a feeling of the turmoil which surrounded the evolution issue. Each man influenced the other's thinking. Because of Darwin, for example, Lyell came to the conclusion that the extinction of species was due to natural causes. He in turn "prepared the way for Darwin's explanation of the origin of species by natural selection" (p. 205). The esteem that Darwin felt for Lyell's important work *Principles of Geology* is evident in a letter to another unnamed colleague, ". . . showed me clearly the wonderful superiority of Lyell's manner of treating geology compared with that of any other author whose work I had with me or ever afterwards read" (p. 116). Bailey, in this biography, shows Lyell as a man of great intellect and varied interests who received many honors and who was knighted in 1848 by Queen Victoria. Lyell published several editions of his *Principles* and his *Elements of Geology*, but, according to the author, his *Antiquity of Man*, ". . . would place him among the foremost of scientific authors. Its scope is amazing" (p. 198). In it he combines geology with archaeology and anthropology bringing to fruition a career that gave the world a more comprehensive geological time table. Bailey's book includes photographs of areas over the world of geologic and archaeologic interest to Lyell and also includes sketches and maps of some of these areas. Other books about Lyell include *Charles Lyell: The Years to 1841* by Leonard G. Wilson (1972) and *The Life, Letters and Journals of Sir Charles Lyell* edited by Katherine M. Lyell (1881).

Clarence King: a Biography. By Thurman Wilkins with the help of Caroline Hinkley. Rev. and enlarged ed. Albuquerque: University of New Mexico Press; 1988. 524 p. Photographs. Bibliography. Notes. Index.

Perhaps the regard that others felt for Clarence King (1842-1901) can best be summed up by an excerpt from a letter written by John Wesley Powell to President Garfield. In it he says, "King was the pioneer and founder of a whole system of survey work which was novel and original, exactly suited to the region and which . . . will on the whole be the only practicable system for the far West. To have grasped the difficulties of the problem at the very outset of the work and to have planned beforehand a system of operations for overcoming them rapidly and easily . . . indicated in King an orderly, sagacious, logical mind which places him among the truly great men of science in the estimation of those who know how to appreciate his work" (p. 226). This support led to King's appointment in 1879 as Director of the U. S. Geological Survey. King's love of science was encouraged by his mother who raised him alone since his father died at a young age. He attended Yale Scientific School, where he later wrote, "An American professor of a classical subject felt entitled . . . to look down upon a teacher of natural science, and by the same quaint sort of logic the Yale academic students excluded 'scientifics' from the boat and fence" (p. 32). He did, however, receive a degree in 1862 completing studies in chemistry, physics, and geology, which included mineralogy, and a special study of crystals. He was greatly influenced by Harvard Professor Louis Agassiz's propoundments on the ice-age theory and came away "more sure than ever that his vocation lay in geology; in that field, his religious, esthetic, and intellectual interests all converged, and he approached the study of the earth as a quest for ethical and esthetic values fully as much as for scientific fact" (p. 42). Author Thurman Wilkins gives an interesting account of King's first trans-continental trip with close friend James T. Gardiner, of King's association with Professor Josiah D. Whitney while working voluntarily with the California Geological Survey, and of King's first field trip in 1863 with William H. Brewer. Many colorful accounts about King's life are related by Wilkins, among them encounters with Indians, stampeding buffaloes, and a grizzly bear. Many were impressed by King's early exploits and endorsed his plan to explore the region that the transcontinental railroad was to span. He was only twenty-five when he was named U. S. Geologist in charge of the Geological Exploration of the Fortieth Parallel. Although the Fortieth Parallel Survey was probably the highlight of King's career, he also made trips into Mexico, Alaska, and Cuba. He also was involved in several mining and other commercial ventures, but evidently was not too successful. Wilkins includes a considerable bibliography and notes section along with a detailed index. He laments, in his preface, that King left no autobiography or personal chronicles, but that details were pieced together from letters, diaries, and field books of his associates.

Fossils and Flies: The Life of a Compleat Scientist Samuel Wendell Williston (1851-1918). By Elizabeth Noble Shor. Norman: University of Oklahoma Press; 1971. 285 p. Illustrations. Photographs. Bibliography.

Elizabeth Shor's biography of Samuel Wendell Williston is an intriguing look at one of the foremost paleontologists and entomologists of the early twentieth century. Shor researched Williston's life by examining family letters, though she says any from Samuel are rare, except for those written from the field to O. C. Marsh who was the United States' first professor of paleontology. "Recollections," Williston's own unpublished record of his life was a major source of material. The result is a thoroughly readable and lively account about this authority on fossil reptiles. Williston had "entered paleontology in 1874 in its rough-and-tumble days" (p. 4) when fossil hunters "vied and fought among themselves . . . determined to gather all the fossils in the West" (p. 3), which "represented one huge prehistoric burial ground" (p. 69). The excerpts from "Recollections" telling of his exploits in Kansas, Wyoming, and Texas are very descriptive and fun to read. For example, he recalls an expedition in north Texas when a cloudburst caused the Little Wichita River to go on a rampage, making it impossible to get to town for provisions. After subsisting on jackrabbits and bullfrogs for a week, "Williston insisted on saying a grace which was:

> Jackrabbits hot, jackrabbits cold,
> Jackrabbits young, jackrabbits old,
> Jackrabbits tender, jackrabbits tough,
> Thank the good Lord we've had enough.

Williston, however, was a serious researcher, as is shown in the ample bibliography of his works, and he evidently loved the work for which he received many honors. He was to wonder, though, in his later years if "what I have done amounts to much . . . too late now for regrets and I know that I have had a good time — many happy hours with bugs and bones." (p. 256)

James Hutton: the Founder of Modern Geology. By Edward Battersby Bailey. Amsterdam: Elsevier; 1967. 161 p.

James Hutton (1726-1797) was enrolled at the University of Edinburgh when a chance remark by a professor of logic piqued an interest in chemical phenomena. He soon changed his course of study from law to medicine since it was the "available subject most akin to chemistry" (p. 2). After completing studies in Edinburgh and Paris, he eventually received his M. D. at Leyden. However, he again changed his mind and turned to agriculture rather than hazard the uncertainties of life as a physician. His friends and associates included economist Adam Smith, chemist Dr. Joseph Black, and mathematician John Playfair. Playfair's part in Hutton's life, in particular, is emphasized because he played such an important role in clarifying and interpreting Hutton's theories. Author Sir Edward Bailey briefly touches on Hutton's early life, and then concentrates on the events surrounding his definitive three volume work, *Theory of the Earth*. Bailey looks at this work chapter by chapter, and interestingly, provides appropriate descriptions of Hutton's experiments along with his field trip notes in order to explain Hutton's deductions. The author also details the controversies that surrounded Hutton because of his assumption that the earth was constantly changing rather than being static. This was in direct contradiction to religious theorists of the day who believed that the earth had been completely formed 6000 years earlier and only catastrophes could change its features. Hutton, however, believed that heat played an important part in the formation of the earth, that changes were brought about by natural processes, and that it would continue to change through these same processes. He is considered the founder of modern geology because he was among the first to examine the earth using experimental methods.

John Milne: Father of Modern Seismology. By A.L. Herbert-Gustar and P.A. Nott. Tenterden, Kent: Paul Norbury Publications; 1980. 198 p. Bibliography.

The authors of this biography of John Milne (1850-1913) started out to "make detailed studies of the lives of prominent scientists and technologists who had been in some way associated with the Isle of Wight" (p. xiv). They started with Milne and found him so fascinating that they went no farther. Milne had studied geology and mining in England, and had made several important expeditions to Iceland, Newfoundland, and the Middle East. In Newfoundland he became interested in the action of ice on rock resulting in the publication of several studies. After returning to England in 1873, Milne grabbed the opportunity to travel to the Middle East with explorer and biblical scholar Dr. Charles Beke, who was in search of the true site of Mount Sinai. The controversy over whether they found the true site raged for several months. Milne took no part in the dispute, for the work he had done was purely geological. Several of the sketches Milne did for Beke are included with the account of the trip. It "proved to be a success for him since he had been able to prepare and have read important papers before the Royal Geographical Society which, in turn, increased his growing reputation as a competent geologist" (p. 19) In 1874 he was appointed Professor of Mining and Geology at the Imperial College of Engineering in Tokyo. One chapter is devoted to Milne's long and arduous overland journey that took him first to Sweden, then across Russia, Siberia, and China to Japan. It took 203 days to get to Tokyo, and oddly enough, on his first night there was an earthquake. It became a "challenge which could not be ignored" (p. 71). He eventually began to study this phenomenon in depth and consequently made it his life's work. Milne was to stay in Japan for twenty years working diligently, and as one Japanese colleague put it ". . . in short, he threatens to exhaust all that there is for workers in seismology to investigate" (p. 85). Among the drawings and photographs that the authors have included, are seismographs that Milne developed over the years. On a different note, Herbert-Gustar and Nott have included an appendix on Milne, "The Storyteller," who evidently wrote fiction under various pseudonyms. They also include several slides with explanatory notes from his collection on Japanese life. Milne was a prolific writer, as the substantial bibliography shows.

Memoirs of an Unrepentant Field Geologist: A Candid Profile of Some Geologists and Their Science, 1921-1981. By F. J. Pettijohn. Chicago: University of Chicago Press; 1984. 260 p. Photographs. References. Index.

Pettijohn's autobiography looks at the field of geology in a time of development and growth. I have included it in this select bibliography mainly because it covers sparse ground.

Although Pettijohn tells briefly about his early life, the real story begins with his study of geology at the University of Minnesota in 1921. It spans the fifty years up until his retirement as Chairman of the Department of Geology at Johns Hopkins University. Pettijohn's chief work was on the Precambrian, with a side-track into sedimentology during his tenure at the University of Chicago. He is very descriptive and either has a wonderful memory, or he kept great notes in a daily journal! He devotes one chapter to a month-long canoe trip from Minnesota up into northwestern Ontario taken in the summer of 1927 with a colleague from Oberlin where he had been teaching. Although he recalls unpleasant episodes, such as being mired knee-deep on a portage or being plagued by flies and mosquitoes, his discovery of unmapped Archean conglomerates became the subject of his research for the next ten years and led to a lifetime study of Precambrian geology. Probably the chief value of Pettijohn's book is the detailed look he takes at the geology departments of Berkeley, Chicago, and Johns Hopkins and how they evolved. He specifically talks about geologists he has known who have made major impacts in the various fields of geology. An index to names, or a "Cast of Characters" as he calls it, is a feature of his book, along with references, or "Memorials and other Historical Materials." He also includes many photographs of people and places.

Powell of the Colorado. By William Culp Darrah. Princeton, NJ: Princeton University Press; 1969. 426 p. Photographs. Bibliography.

William Culp Darrah's biography of John Wesley Powell (1834-1902) is a readable account of the life of a remarkable man who was instrumental in setting cartographic standards for the United States and in furthering a system of geologic nomenclature that is basically still in use. It shows his determination to study science even though finances and family obligations stood in the way. He fought to get proper schooling and, for awhile, taught in order to earn enough to go to college, only to have his education interrupted by the Civil War.

The loss of an arm during the war did not deter him. His energy and curiosity were instrumental in his work to put science to use for the benefit of the people and to make it the basis of progress. Major Powell was a pioneer in encouraging conservation and in developing federal land policy. His trips through the canyons of the Green and Colorado Rivers and years of exploration convinced him of the necessity of an accurate topographic survey of the west. He was appointed Director of the Bureau of Ethnology in 1879 and Director of the Geological Survey in 1881. Powell successfully combined these two responsibilities, carrying on studies of the Indians along with his geologic studies, even though he found it necessary to battle Congress and others who found his ideas too radical. Darrah has captured the life of Powell by collecting and examining nearly 6500 items of Powelliana which, as he notes in his preface, came from many friends, acquaintances, and willing correspondents. "It would be impractical to enumerate more than a careful selection of the principal sources used in this biography—the more so because this study has been based primarily upon original and unpublished sources" (p. 401). The bibliography, however, is an impressive list of sources for those wanting to delve further into Major John Wesley Powell's contribution to the geological and anthropological sciences.

Scientist of Empire: Sir Roderick Murchison, Scientific Exploration and Victorian Imperialism. By Robert A. Stafford. Cambridge: Cambridge University Press; 1989. Bibliography. Index.

Robert Stafford describes Roderick Murchison (1792-1871), the young man, as one who had idle habits and who was more interested in fox hunting than finding an avocation. He "performed poorly in the classroom but excelled at dangerous escapades" (p. 4). Fortunately, he married a woman of intelligence and culture who, when their finances were deteriorating, was able to get him to settle down. From 1816-1818, the Murchisons stayed in Europe where he became interested in geological formations in the Alps. After returning to England, he eventually began studies at the Royal Institution and in 1825 was admitted as a fellow of the Geological Society. He was twice President of the Society, as well as serving in various capacities in other scientific organizations, in particular, the Royal Geographical Society. Murchison is described as being an excellent observer with a capacity for detail who believed that change was unavoidable. His exploratory work in Wales led to the publication in 1838 of *The Silurian System*, which caused a bitter disagreement with fellow geologist Adam Sedgwick over the Silurian-Cambrian stratigraphic boundary. Murchison received several honors from the emperor of Russia for extensive field work in the Russian Baltic provinces, a "geologically little known area" (p. 11), and for completing the first accurate geological map of Russia. Stafford's extensive research provides a sound basis for his analysis of the interplay between science and the expansion of the British Empire in which Murchison had a major role. When he was appointed Director-General of the British Geological Survey, Murchison had the power to promote exploration for commercial purposes in the British Empire and abroad. His influence covered the globe as he "maneuvered to institutionalize natural science as an integral component of both imperial administration and foreign policy" (p. 1). He was opportunistic and fostered economic development, "aiding the search for industrial minerals in Britain while discovering new markets, sources of treasure, and supplies of raw materials abroad" (p. 190). The author's extensive bibliography and notes section would be most useful for those interested in looking further into Murchison's life and British imperialism.

The Stone Lady: A Memoir of Florence Bascom. By Isabel Fothergill Smith. Bryn Mawr College; 1981. 49 p. Photographs. Bibliography.

This is a short biography that looks at the life of geologist Florence Bascom (1862-1945) who was the first female to receive a Ph.D. at Johns Hopkins University. It was written by a long-time friend to whom Bascom was a mentor, and tells of her upbringing by a liberal, intellectual father who encouraged his daughter scholastically in a time that was more attuned to teaching social graces to women. Instead, Bascom was doing geological field work "like a regular feller" (p. 9). Florence Bascom was a woman of several "firsts." She was the first woman member of the U.S. Geological Survey, the first woman member of the Geological Society of America, as well as being its first woman officer. After receiving the Ph.D. in 1893, she taught at Ohio State until 1895 when she accepted Bryn Mawr's offer to join its faculty. She was to stay in the Department of Geology there until her retirement in 1927. She spent 1906 in Germany studying optical crystallography, and after her return to the United States she was instrumental in introducing the new techniques she had learned. Combining her commitment to her academic career with working in government and industry, she was "involved in applying her skill to extending our knowledge of the world" (p. 5). She eventually held the position of full geologist in the petrographical section of the U.S. Geological Survey, where she was involved in the preparation of seven folios on the Appalachian Piedmont. A bibliography of approximately forty articles by Bascom is included in Isabel Smith's tribute to her teacher and friend who was, according to one of her associates in the Geological Survey, "a unique figure in the science of Geology in America" (p. 45).

REFERENCES

1. Adams, Frank Dawson. *The Birth and Development of the Geological Sciences*. New York: Dover Publications; 1954.

2. Porter, Roy. *The Making of Geology: Earth Science in Britain 1660-1815*. Cambridge: Cambridge University Press; 1977.

3. Fenton, Carroll, Lane Fenton and Mildred Adams Fenton. *Giants of Geology*. Garden City, NY: Doubleday; 1956.

4. Faul, Henry and Carol Faul. *It Began With a Stone: A History of Geology From the Stone Age to the Age of Plate Tectonics*. New York: John Wiley and Sons; 1983.

A Selection of Biographies
of Animal Scientists

Flora Cobb

INTRODUCTION

Emphasis in education is being placed on the need for people to be able to think critically in order that they can assess and use the flood of information available on almost any topic. Another current concern is that fewer students from the U.S. are choosing to pursue graduate study or careers in science. It seems that some exploration of scientists' lives might inspire individual students to follow any leanings toward science they may have. The biographies of animal scientists described here were chosen because they attempt to show the reader the whole personality of the biographee including contributions made to his/her field. Obviously, the titles included represent only a tiny selection of available biographies. Readability by a wide audience and a tone informal enough to make the personality of each biographee accessible were the primary criteria used in deciding which titles to include in this selection. Incidents and thought processes are shared with the readers of these volumes making clear how broad-minded, persevering, and flexible good scientists tend to be.

Flora Cobb is Science Reference Librarian at Virginia Polytechnic Institute and State University, Blacksburg, VA 24061-0434. She holds a Bachelor's degree in Zoology from the University of Tennessee and an MLS from the University of Texas, Austin.

121

Through a Window: My Thirty Years with the Chimpanzees of Gombe. By Jane Goodall. Boston: Houghton Mifflin; 1990. 268 p. ISBN 0395500818.

This work may seem to be more a biography of chimpanzees than of Jane Goodall, but the stories of chimpanzees Dr. Goodall has studied during her thirty years in Zaire are good learning tools, especially for people who may otherwise have no exposure to chimpanzees. The stories express Dr. Goodall's life focus and work. Anecdotes about the chimpanzees and life histories of prominent individuals and groups are the major focus of the text. With a clear writing style, Dr. Goodall incorporates anecdotes, observations, and their scientific implications and interpretations with observations about her own life and emotions.

Early chapters describe how Jane became involved in chimpanzee studies and how she began work in Gombe. Initially Jane had problems with her working and reporting style not being perceived to be "scientific" enough. Her tendency to anthropomorphize the chimps was criticized rather than being viewed as fundamental to the nature of her work. Bits of information about chimpanzees' behavioral patterns and cultural interactions are delivered throughout the book, usually as a summary to some narrative about events in a particular chimp group. Similarities between chimpanzees and humans are discussed frequently. Observations of baboons are included as well, with discussions of baboons' and chimps' interactions and comments on the differences between the two. Chapters are devoted to individual chimpanzees Jane observed. Some chapters focus on emotions or behaviors characteristic of chimps and observed in the chimp groups over the years. Transitions between subjects and groups are smooth.

Two appendices appear at the end of the book; one is on exploitation of non-human animals and another offers details of the state of the chimpanzee conservation movement and sanctuaries. Jane decries the treatment many animals receive in laboratories where they are used for scientific experimentation. She proposes that ethics with regard to humans' use of non-human animals need to take into consideration animals' capacities for feeling. Jane suggests that more stringent standards for defining what makes research essential and for improving living situations for research animals are needed. The book's index is helpful—names of individual chimps and baboons are indexed. A map of sanctuaries and conservation areas is provided with the density of chimp populations indicated.

Woman in the Mists: the Story of Dian Fossey and the Mountain Gorillas of Africa. By Farley Mowat. New York City: Warner Books; 1987. 380 p. ISBN 0446513601.

Farley Mowat presents an account of Dian Fossey's life with depth and sensitivity in this biography based on papers, letters, and journals as well as interviews with coworkers, students, relatives, and others who knew Dian. Passages taken directly from primary source material are printed in bold type throughout the text; Mowat weaves his own narrative around such passages to describe events and scenes, people and gorillas. Mowat explains Dian's forceful personality and strong drive as he presents the events of her life leading up to and during her time in Africa. Conflicts and difficulties Dian faced on many levels—political, cultural, physical, emotional, academic—are depicted in a way that makes her passion for understanding and protecting the gorillas come alive for the reader. Although many people might judge Dian as having been too uncompromising and extreme in her demands, the book makes clear that a person of less strength of character (some would say eccentricity) probably would not have been able to accomplish as much as Dian did.

Maps of the Karisoke research Center area and mountain areas where the gorilla groups lived are provided. Photographs and a good index enhance the book.

The Dark Romance of Dian Fossey. By Harold Hayes. New York City: Simon and Schuster; 1990. 351 pp. ISBN 0701133147.

This biography is a novel-like account of Dian Fossey's life and work with the gorillas in Africa. The facts and many of the stories are the same as those presented in the biography by Mowat, but Hayes uses his own narrative without passages taken directly from Dian's journals and other primary sources, though he did rely on such sources heavily in writing the book, and occasional quotes from journals and letters appear. The writing style is very engaging and readable. Hayes did extensive research which included listening to tapes of lectures and interviews and conducting his own interviews of people in Dian's life. A long bibliography lists books, articles, theses, lectures, and interviews which served as Hayes's references.

Development of Dian's early life—experience with her mother and stepfather, influences of friends in Kentucky—all seem to be complete and fit nicely in the exploration of her character. Hayes's descriptions and deductions regarding Fossey's personality and feelings seem very likely to be accurate, and they help the reader to get a feel for Dian's personality and drive. Stories of some times in Dian's life such as the time she spent at Cornell University lecturing and writing (and rewriting) her book, *Gorillas in the Mist*, serve to show the breadth of her experience and to reinforce the reader's sense of her personality.

More complete details about the tragic end of Fossey's life are given than in Mowat's book. The scandal shrouding her death and the accusation of a student named Wayne McGuire are described very clearly.

Overall the book's tone is not as intense as that of Mowat's, and the writing integrates very smoothly characters and times of Fossey's life. Impressions Fossey's friends, relatives, and associates have about her and her work are presented. An excellent index and photographs enhance the story.

No Woman Tenderfoot: Florence Merriam Bailey, Pioneer Naturalist. By Harriett Kofalk. College Station, Texas: Texas A&M Univ. Press; 1989. 225 p. ISBN 0890963789. Price $19.95.

This biography of Florence Merriam Bailey, an early female bird-specialist, explores Ms. Bailey's quest to make the study and appreciation of birds in any locale appealing as a satisfying pastime. Bailey strived to make the scientific aspects of studying birds accessible to people of all ages and types. She is likely not well known to non-ornithologists, and this book offers an excellent opportunity to be exposed to the life and work of a scientist who is exceptional in many ways, one being that she was one of few women to become active in a scientific field in the late nineteenth and early twentieth centuries. Florence's life was imbued early with a pattern of discovery and outdoor activity as she grew up in northern New York state and later attended Smith College. Kofalk fully develops the details of the influence Florence's family and husband had on her work in ornithology. Florence's brother, C. Hart Merriam, worked for the U.S. Biological Survey (predecessor of the Fisheries and Wildlife Service), and Florence married a co-worker of his, Vernon Bailey, a mammalogist.

Florence pursued migration and identification which fueled her advocacy for appreciation of birds in their natural habitats. Accompanying her husband on Biological Survey expeditions, Florence traveled around the United States observing and reporting on birds she saw in diverse areas.

Florence led a campaign against hunting for the sake of killing. She offered strong support to organizations which worked to instill in sportsmen an appreciation for the animals and birds they hunted, and she wrote many articles on that topic for publications aimed at sportsmen. A related fervent fight Florence carried out was against the rampant and extravagant use of feathers in ornamentation of women's hats. She became extremely active in the Audubon Society and the American Ornithologists' Union and used those affiliations as a vehicle for promoting education and public interest.

Florence was a prolific writer throughout her life. Articles she wrote appeared in *Audubon* and other naturalist magazines, and she wrote numerous books, handbooks, and identification guides. Her *Handbook of Birds of the Western United States* served as a standard bird identification manual for decades and underwent at least four revisions. Skilled at describing field experiences, Florence could use records of her observations to assist other birders in recognizing bird species. Florence was gifted at combining scientific information with the personal quality of her observations.

In the book's introduction the author comments on the idea that exploring history is valuable because it helps people to gain appreciation for the present. Examination of Florence Bailey's life is decidedly valuable as an opportunity to learn how a female ornithologist in the late 19th and early 20th centuries challenged the expectations of what a woman's life could be like, expanded her horizons to accomplish what she found to be satisfying, and shared her passion for birds with others. Reading this story of Florence's life will enrich the work of anyone pursuing/reviewing her writings.

Dance of the Wolves. By Roger Peters. New York: McGraw Hill; 1986. 222 p. ISBN 0070495807.

The goal of this book is not to serve as a scientific treatise but to share the knowledge of the behavior of wolves with the general public, the author states in the introduction. Peters accomplishes his goal and shares also a sense of the emotions and of the extremes of self-doubt and satisfaction which seem to be common to anyone undertaking a risk to try something which is at once demanding and wonderful. Exploring the unknown, in both geographical territory and personal experience, and the self-knowledge which can come from such exploration are the main themes of this book.

Detailed explanation of ways researchers track wolves is fascinating. Wolves are labelled by means of radio transmitters attached to collars the wolves wear. The wolves are trapped to give scientists a chance to attach the collars. Flying in an airplane over the wolves' territory, scientists can use radio receivers to locate the radio-collared wolves and track their movements. Tracking wolves on the ground involves studying and following scat, urine, tracks, broken twigs, and any other evidence of the wolves' presence.

Peters came to feel very attached to the wolves he observed, and through his descriptions of the wolf groups, the reader is made to feel closely involved with the animals. Interesting discoveries and knowledge of wolves are shared throughout the book. For example, Peters and his colleagues learned that, although wolves most commonly use scent and rote learning of trails to find their way around, sometimes they do not use those skills. Rather, a more direct route to a destination might be followed by the wolves divining the way using some kind of direction detecting ability and a special awareness of spatial relationships. It was this phenomenon which Peters was studying for his dissertation.

The writing style is not particularly good; repetitive syntax and drab reporting style detract from the subject matter and leave the book interesting but not enthralling. Terminology used is described in the text only once, and there is no glossary or index to help enhance the context in which such terms are used.

Nevertheless, the scenes and work described are exciting to anyone who likes wolves and the outdoors. Photos are provided which make the subjects seem more real to the reader.

Konrad Lorenz. By Alec Nisbett. New York City: Harcourt, Brace, Jovanovich; 1977. 240 p. ISBN 0-312-03901-8.

On Life and Living: Konrad Lorenz in conversation with Kurt Mundl. By Kurt Mundl. New York City: St. Martin's Press, Inc.; 1990. 166 p. ISBN 0-15-147286-6.

Together these two books provide a view of Konrad Lorenz's life, work, and character through the telling of fascinating tales and by raising questions of concern to all who ponder humans' place in the world and our relationship to animals.

Both books chronicle the events of Lorenz's life and his efforts in the establishment of research and teaching in ethology. Nisbett's book is written in traditional biographical style, while Mundl's is a collection of transcribed interviews with Lorenz which have been translated from German.

Nisbett states in the introduction to his book that his purpose in writing it is to aid people in discovering and understanding Lorenz's life, including his many ideas, scientific contributions, and writings by and about him. Nisbett successfully accomplishes his goal by providing details of Lorenz's work and alliances and by including anecdotes, quotes, and personal stories which make for lively reading.

Animal behavior as a discipline of study has had different labels and emphases in America from those in Europe. In America behaviorism traditionally emphasized the role of learning, while in Europe ethology was more concerned with instinct. Differences between the ideas of instinct and learning definitely exist in Lorenz's mind. Many of his ideas about ethology were long resisted in America because they support belief in the existence of instincts and pre-programmed individual capacities. Resistance stemmed from a belief throughout American history in equality among people and the in idea that opportunity combined with individual effort are what distinguish people in society. Gaps extant in the description or understanding of organisms' development and behavior seem to have been held sacrosanct by some scientists because they allow a logical place for soul, free will, or consciousness. Lorenz can not abide by that idea, and he believes some gaps in knowledge exist simply because scientists do not know the answer.

Lorenz's early life, education, family life, experience during World War II as a Russian prisoner of war, and scholarly achievements are presented in detail.

Kurt Mundl's book of interviews with Konrad Lorenz, originally published in German in 1988, is both delightful and challenging because lofty questions are examined in a familiar tone. Lorenz speaks openly in clear language about his life and work. Anecdotes about experiences with animals make the book likely to appeal to younger or casual readers; Lorenz tells one story about lying down atop a Kimodo dragon. He advocates allowing children to have an aquarium, at the very least, as an opportunity to instill in them a respect for and curiosity about animals. Lorenz considers awareness of the natural world to be fundamentally necessary to a satisfying life. The book is divided into four major sections having the following titles: Excerpts from the Life of a Natural Scientist; Observations on Living Systems; The Misuse of Creation; Man and His World. Each section contains short chapters, and each chapter is Lorenz's answer to a question posed by the interviewer.

The challenge presented by this book comes from Lorenz's heartfelt concern over the dilemmas created by society's use of science as a tool which can make life more comfortable but which may contribute also to ruination of natural beauty. In addition to comments on his own life and work, Lorenz presents his views on some problems in society such as conservation, ecological awareness, nature preservation, use of animals in scientific experimentation, and education. Personal opinions expressed about the fate of the earth, raising children, and the future all reinforce an underlying theme that it is necessary for people to take responsibility for human action. Lorenz blames overpopulation, in part, for the increasing occurrence of aggression and a lack of warmth among people; such phenomena, he believes, represent the beginning of inhumanity. People have become slaves to comfort and continue to force nature out of daily activities and lifestyles. Lorenz thinks such behavior is leading to a dead end for the human species. These are heavy thoughts coming from one who served as "mother" to numerous baby birds and other creatures in the course of his studies on imprinting in animal behavior.

A side benefit of reading this book is that Lorenz refers to books and films that have influenced him or that he enjoyed (some have not been translated into English). The tone of the book is much more that of a personal discussion rather than a chronology of events and accomplishments.

Leaders in the Study of Animal Behavior: Autobiographical Perspectives. Ed. by Donald A. Dewsbury. London: Bucknell University Press through Associated University Presses, Inc.; 1985. 512 p. ISBN

Nineteen short autobiographies by "distinguished animal behaviorists" whose work spans the twentieth century make up this volume. Two strengths of this work are its presentation of the emergence and underlying philosophy of the discipline of ethology and the excellent opportunity it provides to view the lives of scientists through the eyes of the scientists themselves. Exceptionally clear writing style is a pervasive characteristic of the pieces in this volume making complicated concepts accessible to readers who may not know much about behavior studies. In almost all cases the author presents a compact chronology of his life highlighting experiences and relationships with people and/or animals which were pivotal to his path through a life in science. Most of the authors are still alive, and many comment on their continuing interests and immediate research goals.

In each capsule view the scientist successfully conveys the theoretical or philosophical struggles surrounding his work. The contexts for specific experiments and ideas are made clear to the reader together with an explanation of the author's sense of the significance of the questions under investigation.

For sources of inspiration about what a life working in science (ethology, at least) might be like, this volume offers a feast of possibilities. Choices made, chances taken, and strokes of good luck are divulged by each of the ethologists who contributed his autobiography for this collection. Lots of interrelationships between the authors of the pieces emerge; they have known each other in the roles of pupil, teacher, and colleague. Similarly, cooperation among this group of ethologists and others in the field is evident from the references listed after each chapter in which most of the authors include both works which they have written and which have been influential.

Animal Days. By Desmond Morris. New York City: William Morrow and Company, Inc.; 1980. 304 p. ISBN 0-688-03666-X.

Animal Days tells the story of Desmond Morris's life, beginning as a student of animal behavior and leading to posts as curator of mammals at the London Zoo and as director of London's Institute of Contemporary Art. As a student at Oxford, Morris worked with zebra finches with the hope that he could use what he learned about their behavior differences as a key to explain the evolution of the whole finch family. Eventually, Morris studied under Niko Tinbergen, Nobel prize winning ethologist (Tinbergen's autobiography is included in the collection entitled, *Leaders in the Study of Animal Behavior*). Tinbergen is well known for his work on instincts, and Morris credits him with introducing his students to field experimentation methods in which the observational powers of the naturalist are combined with (and granted scientific validity by) scientific thinking processes of a lab worker. During his tenure as a graduate student, Morris had the opportunity to be involved in the first International Ethological Conferences which began as small, informal meetings where the atmosphere fostered lively exchange among scientists discussing problems they were currently studying.

While Desmond Morris pursued his graduate degree, his wife, Ramona, also trained in zoology, undertook bird studies, including training crows to talk. The Morrises worked together to publish several books about their broad experiences and observations. Morris worked in the television unit of the London Zoo after finishing graduate school, and while there he helped to create a weekly television program designed to educate the public and generate awareness of animals. The show, called "Zootime," was a broadcast showcase for zoo inhabitants and discoveries about animal behavior. Morris later became the London Zoo's Curator of Mammals. In *Animal Days* many unusual and, in retrospect, humorous tales are related about Morris's experiences while working as curator, such as being called out to examine a beached whale which had turned up in the river Thames.

The feature story of *Animal Days* is that of Congo, a chimpanzee, who learned to draw and paint with various media. Working with Congo gave Morris the opportunity to pursue his interest in art and its utility as a vehicle for studying human nature and behavior. The chance to serve as director of the Institute of Contemporary Art came about as a result of Morris's interest in art and administrative experience he gained while working for the zoo.

A strength of this book is the emphasis placed on the importance of serendipity to success; Morris makes the value of the unexpected evident through the anecdotes related to his scientific investigations over the years.

Courageous Cattlemen. By Robert C. deBaca. Huxley, Iowa: R.C. de Baca (Champaign, IL: KOWA Graphics); 1990. 357 p. ISBN 0813806186.

This volume, comprising sixty-three chapters on individuals and major concepts, is a chronicle of the growth of the cattle breeding industry in the United States. The subjects of the chapters are known for their innovations and cooperation in the breeding and improvement of beef cattle in the twentieth century. Animal scientists in the traditional roles of research and teaching are included together with cattle breeders who have had exceptional success in applying principles of animal husbandry. De Baca says research has transformed animal breeding from an art into a science. The people portrayed in this book demonstrate how research and practical knowledge of traditions in breeding have been integrated.

In the book's preface de Baca sets the stage nicely for the portraits which follow by discussing the evolution of the understanding of genetics, contrasting the early years of animal breeding with the tremendous technological advances of the present day. Agricultural extension workers and land-grant university researchers are shown to have played a crucial role in the process of sharing results of research with farmers who can apply the information. De Baca laments the bias which agricultural industrialists have given to commentary over the years making the themes of many publications seem to be that of a sales pitch rather than that of a channel for exchange of information. Agricultural journalists writing for the popular press have been helpful in advocating trends desirable both to consumers and to producers, such as the need for lean, fast-growing cattle. Unfortunately, de Baca says, the press often does not credit breeders and researchers who initiate work required to develop the means to meet demands of market trends. *Courageous Cattlemen* is educational because it describes the work of many innovative people who have been willing to take risks and to communicate their findings.

A few of the chapters discuss important associations for cattle breeding and improvement, research centers and foundations, and breeding methods and principles. These chapters show how the people presented in this book have worked together to standardize successful techniques and to share their knowledge and experience.

Adventures of a Zoologist. By Victor B. Scheffer. New York: Charles Scribner's Sons; 1980. 204 p. ISBN 0-684-16439-6.

Victor Scheffer tells an exciting story of his life as a zoologist specializing in the study of sea mammals and becoming involved in protection of wildlife. Scheffer's studies and first work experience took place in Washington, where he later taught wildlife classes at the university in Seattle. The bulk of Scheffer's career was spent as a sea mammal specialist for the U.S. Fish and Wildlife Service. Work with fur, harbor, and harp seals presented opportunities for Scheffer to become involved with determining wildlife protection policy to prevent killing of endangered species for luxury products or sport. A prolific author of scientific articles, Scheffer also enjoyed success as an author of several books. His writing reflects his determination to make scientific thought and discoveries accessible to the public.

Tales about trips to Alaska's Aleutian islands, the Cayman islands, New Zealand, and other exotic locations are rich with anecdotes about coworkers and unexpected scientific discoveries made along the way. The thrill of covering new territory and cataloging its resident species is made vivid to the reader. In some cases Scheffer and colleagues discovered populations of species which were previously thought to be extinct.

Scheffer writes with a balanced and enlightened view about complex moral, ethical, and psychological issues associated with judgments about animal welfare and protection. Thoughtful observations about progress in his career and his development as a zoologist are shared in this autobiography. Scheffer admits to some actions he calls "poor zoology" in which he claims to have been impetuous. Looking back, Scheffer understands why some actions failed because they were contrary to the natural process and may even have caused harm to some individual animals. Such open examination of his "maturation" as a scientist bridges the gap between science and popular knowledge. Scheffer translates scientific jargon to the vernacular, and his writing advocates his view of the wholeness of life and respect for nature.

Aldo Leopold: His Life and Work. By Curt Meine. Madison: Univ. of Wisconsin Press; 1987. 638 p. ISBN 0299114902.

Aldo Leopold is known as an early forester and naturalist who proposed bold ideas for conservation. Justification for including this biography of Leopold by Curt Meine with books about animal scientists comes from Leopold's work with wildlife. His book, *Game Management*, was among the first attempts to define methods for preserving game through appropriate use of land and its resources. *Game Management* continues to be used among wildlife educators today. Leopold is best known to the public for his writing, such as *Sand County Almanac*, and this book fills a need to make known the story of the rest of his life.

Meine's biography of Leopold offers an excellent introduction to the beginnings of forestry and conservation in the United

States as well as details of Leopold's life. Chronicles of Leopold's ancestors who settled in the U.S. midwest set the stage for understanding the world into which Leopold was born. Quotations from interviews, letters, and other primary sources help the reader to come to know Leopold's personality. Naturally talented as a writer, Leopold wrote countless letters home and notes in journals. Passages from such writings offer the reader firsthand accounts of thoughts and events Leopold recorded which indicate the nature and flow of his thoughts.

Meine has done an excellent job of combining Leopold's own work and writing with accounts from family members and coworkers. The resulting details and personality analysis make an intriguing study about guiding principles and ethical issues which may be encountered by anyone interested in conservation. During Leopold's lifetime National Forests were established in the U.S, the Forest Service was created, and training for forestry as a career became available. Leopold grappled with dilemmas about how to reconcile conflicting needs to preserve and to use land and its associated resources; his ideas paralleled struggles to determine policy and appropriate practices at the national level.

Meine carefully presents Leopold's life-long effort to define conservation better. Conflicting desires and interests, as in the case of land use and preservation, Leopold experienced were constantly undergirded by a deep appreciation and respect for nature. The notes and bibliography at the end of this book enhance its value and offer suggestions for further reading.

Plant Breeders and Plant Geneticists: Biographical Information

Dena Rae Thomas

Only within the last 10,000 years have men and women relied on agriculture and the domestication of animals in order to make a living.[1] One can guess that during most of that time humans have tried, through various means, to influence the forms of the plants and animals kept for sustenance, power, companionship, and decoration. From the first time a farmer noticed a plant bearing larger or earlier fruit and tried to increase the occurrence of that plant, we have had avid plant enthusiasts among us.

The individuals selected are from varied backgrounds, education levels, and nationalities, but all shared a fascination with gardens, plants, and the elements of life. Most of the six were born in a fifty-year period, from the middle of the nineteenth century to the beginning of the twentieth.

In reading through the biographies, I hope the variety and the drama of their lives, and their passion for knowledge is conveyed. For each individual, I include bibliographic information on several of their most important written works, as well as citations for and annotations of works about them. Where possible, I have attempted to include sources providing various perspectives on the scientists and their lives.

Dena Rae Thomas is Patents and Instruction Librarian in the Centennial Science & Engineering Library at University of New Mexico, Albuquerque, NM 87131-1466. She holds a Bachelor's degree in Genetics from the University of California at Davis and a Master's in Librarianship from the University of Washington.

Acknowledgement: *Biographical Encyclopedia of Scientists*, John Daintith, ed. New York: Facts on File Inc., 1981 was used throughout in outlining the biographical essays.

LUTHER BURBANK

Born in Lancaster, Massachusetts, on March 7, 1849, Luther Burbank was a renowned American plant breeder. The family farm provided him with his early interest in growing plants, and the town library provided him with materials on current scientific thinking. He read *Variation of Animals and Plants under Domestication* by Charles Darwin, a book which was a leading influence on his life's work. He attended public schools and then the Lancaster Academy.

At the age of twenty, Burbank bought a small parcel of land near Lunenburg, Massachusetts, and for the next five years worked at breeding crops for the local market. During this time, he developed the Burbank potato, one of his most significant achievements. In 1875 he traveled by train to Santa Rosa, California, where his brother Alfred had taken up residence. Initially, Burbank worked as a carpenter to support himself. His mother and sister moved west to Santa Rosa in 1877, and he rented land from his mother to begin his nursery.

Burbank aimed to develop improved varieties of cultivated plants, including trees, flowers, fruits and vegetables. His business began slowly but grew steadily as he reinvested the earnings and his reputation increased. Hybridization and selective breeding allowed him to produce thousands of improved plants. Besides the Burbank potato (which is still in use today), he produced varieties of plums and prunes, apples, blackberries, and cherries, and flowers such as the Shasta daisy, various lilies, cannas and dahlias, among others. He bred a spineless cactus which he hoped would open desert areas to stock raising, but the plants were not used on a wide scale.

Although Burbank's work interested some famous scientists of the time, among them Hugo de Vries, a Dutch geneticist, Burbank was not interested in elucidating the scientific basis of his outstanding results. Indeed, though he knew of Mendel's work, he seems to have been convinced of the possibility of inheritance of acquired characters. In addition to the many new varieties of plants he produced, Burbank served as an inspiration for development and innovation in plant breeding. He died in Santa Rosa on April 11, 1926, after fifty years of work in his gardens and nurseries.

Selected Works by Burbank

Luther Burbank, His Methods and Discoveries, 12 vol. Santa Rosa: The Luther Burbank Society, (1914-15).

How Plants Are Trained to Work for Man, 8 vol. New York: P.F. Collier & Son Company, 1921.

The Harvest of the Years, (an autobiography) with Wilbur Hall. Boston: Houghton Mifflin Company, 1927.

Selected Works About Burbank

de Vries, Hugo. "The production of Horticultural Varieties by Luther Burbank," in *Plant-Breeding*. Chicago: Open Court Publishing Co., 1907.

> De Vries was a fond admirer of Burbank's work, and this essay is a review of his life, his methods, and mainly, his results. Though written by this eminent scientist at the turn of the century, the work is neither technical nor antiquated in tone.

Dreyer, Peter. *"A Gardener Touched with Genius: The Life of Luther Burbank."* Berkeley: University of California Press, 1985.

> An interesting and offbeat presentation of Burbank and his work. Chapters are each built around a theme and though these serve to further the story, they read also as eclectic stand-alone essays. "The Winds of Biology" describes the development of biological thought from Lamarck to Burbank, and stresses, " Burbank's genius was practical and intuitive, not theoretical." "Burbank's Legacy" is more than a list of the famous varieties Burbank created. It shows his influence as a source of inspiration, and also presents some of the reasons that Burbank was shunned by most of the scientists of his time.

GEORGE WASHINGTON CARVER

George Washington Carver, American agricultural chemist, was born in Diamond Grove, Missouri, probably in 1865. He was one of two children of Mary, a slave woman owned by Moses and Susan Carver. Near the end of the Civil War, Mary and the young George were kidnapped. Though the boy was recovered by a neighbor (who, some say, received a racehorse as a reward), Mary was not found, and George and his elder brother Jim were raised in the Carver house as though they were the Carvers' own.

George left the Carver family when he was only about twelve, traveling to a nearby town that had a black school. His thirst for knowledge led him from Missouri to Kansas and to Iowa, a quest lasting the next thirty years. By the late 1880's living in Iowa, he enrolled at Simpson, a small Methodist college, where he developed musical and artistic abilities. His art teacher recognized talent, but worried about the future of a black male painter, and knowing of his love for plants, she suggested that he enroll at Iowa State College, the agricultural school in Ames.

In 1891, as the only black on campus, Carver experienced prejudice and indignities. But he was experienced in living among whites, and soon impressed faculty and students alike with his interest in botany and his talents in art and music. Always a religious person, he attended and taught Bible meetings. His abilities in plant raising, crossbreeding, and grafting were all noted by the faculty. Upon completing his Master's degree in 1896, he received several offers of employment, deciding to accept one from Tuskegee Normal and Industrial Institute in Alabama. The school's director, Booker T. Washington, was known for his leadership in the black community, and for his ideals of self-help and interracial cooperation. Carver accepted the post at Tuskegee as director of a new Agricultural Department and a new experimental station, seeing the position as an effective agent to work for the betterment of his race.

For the next forty-plus years Carver worked at the Institute. His relationship with Washington was often stormy, as the two had widely different outlooks. Carver had a backbreaking schedule of teaching, running research out of an ill-equipped lab, directing experiments on the small experiment station's plots, and doing consulting work, helping nearly anyone who asked in identification of crop or soil problems. The Institute also ran extension programs, sponsoring Farmers' Conferences, and summer schools for teachers.

Noting the severe overproduction of cotton, and its depletion of the soil, Carver had a consistent career desire to end the reliance on King Cotton. He was concerned with the poorest people, whether or not they owned their land, and researched techniques to enable them to turn a profit despite small farms and rudimentary equipment. He continually experimented with alternative crops to cotton, looking for those that would replenish the soil as they gave the farmer something useful. His research centered on three crops: cowpeas, sweet potatoes, and peanuts, all of which, he felt, could help improve the diet of the farmer while replenishing the soil from its debilitating hostage to cotton.

Carver received many awards of recognition for his work: he was elected to the advisory board of the National Agricultural Society in 1916, and also was made a fellow of the Royal Society for the Arts in London in the same year. He received honorary doctorates of science from Simpson College, Selma University, and the University of Rochester. He received the Spingarn Medal of the NAACP in 1923. He was made an honorary member in the American Inventors Society in 1941, a member of the Laureate Chapter of Kappa Delta Pi, an honorary education society, and awarded a fellowship from the Thomas A. Edison Institute.

Carver's value as a symbolic figure probably overrode his contributions as a scientist. He was an effective spokesman and role model for blacks seeking greater control over their destinies. He was a gifted and persuasive speaker, and his speeches had an inspirational quality which affected and empowered audiences, whether they were students searching for a place in the world, industrialists hoping for a renaissance of the South, or farmers attempting to eke out a living from exhausted soil. Carver died on January 5, 1943, leaving the George Washington Carver Museum and Foundation in Tuskegee for displaying and continuing his work.

Works by Carver:

Much of Carver's research is reported in the *Agricultural Experiment Station Bulletin* of the Tuskegee Institute. Selected titles and their numbers and dates follow.

Experiments with sweet potatoes, no. 2, 1898.

The pickling and curing of meat in hot weather, no. 24, 1912.

How to grow the cowpea and 40 ways of preparing it as a table delicacy, no. 35, 1917.

How to make sweet potato flour, starch, sugar, bread, and mock coconut, no. 37, 1918.

How to grow the peanut and 105 ways of preparing it for human consumption. 1942.

Works About Carver:

Many more works exist than are included here. Most examined were either of the sentimental variety or were very dated in presentation.

Edwards, Ethel. *Carver of Tuskegee*. Cincinnati: Psyche Press, 1971.

An affectionate, interpretive biography, this book is accessible to a wide readership. The science is described at a general level. Episodes of Carver's celebrity are described, including a meeting with Franklin Delano Roosevelt, correspondence with Ghandi, and his friendship with Henry Ford.

McMurry, Linda O. *George Washington Carver: Scientist and Symbol*. New York: Oxford University Press, 1981.

This book is a well-researched and scholarly examination of Carver and his work. The author does a good job of explaining how Carver's religious understanding of man as a part of a whole formed the basis for his research and the life path he chose. Though Carver was recognized from the time he graduated from Iowa State as talented in the areas of art, plant breeding, and mycology, he chose to work in Tuskegee, where, he felt, his work would have the most considerable impact among the "fartherest down" — the poor black farmers. The author also argues that Carver's chief contribution was as an interpreter between the scientist and the layman, making scientific progress accessible and understandable to the masses.

HUGO DE VRIES

Hugo de Vries, Dutch botanist and geneticist, was born in Haarlem, Netherlands, on February 16, 1848. He studied botany at Leiden, Heidelberg, and Wurzburg, and became a professor of botany at the University of Amsterdam in 1878. His original interest was plant classification, and later he turned to physiology and evolution. Some of his early studies investigated the water in plant cells, and his work was instrumental in Jacobus Van't Hoff's theory of osmosis. By the 1880's his interest had shifted to inheritance, and in *Intracellular Pangenesis*, (1889) he claimed that nuclear elements, which he called "pangenes" were the determinants of inherited traits.

He conducted plant breeding experiments and by 1896 had observed that offspring of crosses showed a ratio of 3:1, but at that time he did not publish his results. In 1900, he read the same findings in Mendel's work (which had been published in 1864), and hastened to publish his "rediscovery." Both Karl Correns and Erich von Tschermak also published papers in 1900 that reiterated Mendel's work.

De Vries' most influential work, *The Mutation Theory*, (1901-03) was based on his experimental work with the evening primrose, *Oenothera lamarckiana*. The work with *Oenothera* suggested to him that new species arise through mutation. Accordingly, he theorized, evolution occurs as natural selection acts on these large, discontinuous variations rather than on the continuous variation suggested by Darwin and others. De Vries hypothesis helped to explain a number of difficulties in Darwin's theory of evolution, and the mutation theory held many biologists in thrall for the first decade of the 1900's.

By 1915, many had abandoned the mutation theory due to a variety of problems, not the least of which was that de Vries' Oenothera were shown to be polyploids, thus exhibiting a misleading impression of the kind and variety of mutations. De Vries remained a staunch defendant of the mutation theory until the end of his life. He was a professor of botany at Amsterdam for forty years, until 1918, and continued working in plant breeding subsequent to his retirement from the University. Though he is remembered mainly as one of the three "rediscoverers" of Mendel's laws, his most significant contribution to science was in demonstrating that evolution could be studied in an experimental environment. Until his time, most evolutionary thought had arisen from comparative work and inference.

Selected Works by de Vries

Intracellulare Pangenesis. Jena: Gustav Fischer, 1899. (English translation: C.S. Jaeger as *Intracellular Pangenesis*. Chicago: Open Court Publishing Co., 1910.)

Die Mutationstheorie. Leipzig: Von Veit and Co., 1901-1903. 2v. (English translation: J.B. Farmer and A.D. Darbishire as *The Mutation Theory*. Chicago: Open Court Publishing Co., 1909-10.)

Species and Varieties, their Origin by Mutation. Chicago: Open Court Publishing Co., 1905.

Plant-Breeding: Comments on the Experiments of Nillson and Burbank. Chicago: Open Court Publishing Co., 1907.

Selected Works About de Vries

Allen, Garland E. "Hugo de Vries and the Reception of the 'Mutation Theory.'" *Journal of the History of Biology* 2: 55-87 (1969).

> Fascinating explanation of the popularity of de Vries' mutation theory at the turn of the century. Although his supporters

were from different branches of biology and had differing reasons for favoring his theory, de Vries' theory appealed to a dissatisfaction with Darwinism which was then prevalent. The theory accounted for some of the strong faults in natural selection by providing an explanation for the source of variation. The mutation theory, with its discontinuous variation, provided a better fit with number of species and amount of geologic time. Last, the mutation theory was popular because it demonstrated that evolution could be subject to experimental study.

van der Pas, Peter W. "The Correspondence of Hugo de Vries and Charles Darwin," *Janus* 57: 173-213 (1970).

The article begins with a condensed history of Darwin's publications, and describes his influence on de Vries' early work in plant physiology. The two began corresponding in 1875, when Darwin sent de Vries a copy of his book on climbing plants. Their only meeting, in 1878, is described in several of de Vries letters to his friends and relatives. Though only a few of Darwin's letters are included, the article is a fascinating example of scientific progress through correspondence. As an interesting sidebar, there is a short discussion of Johannsen's choice of the word "gene" which he uses in preference to de Vries' use of "pangene" as homage to Darwin.

WILHELM LUDWIG JOHANNSEN

Wilhelm Johannsen, Danish botanist and geneticist, was born in Copenhagen on February 3, 1857. His father was a Danish army officer. The maternal side of his family was interested in German language and culture and Johannsen was accomplished at several languages while still very young. Although he passed the qualifying examination, the family did not have means enough to support a university education, and Johannsen was apprenticed to a pharmacist in 1872. For the next seven years he worked as a pharmacist in Denmark and in Germany, studying chemistry and botany on his own.

Returning to Denmark in 1879, he passed the pharmacist's examination and continued to study chemistry and botany. He became an assistant at the chemistry laboratory in the newly-opened Carlsburg Laboratory under Johand Kjeldahl. He researched various metabolic processes in plants, particularly in seeds, tubers, and buds.

In 1892 Johannsen became a lecturer, and in 1903, professor of botany and plant physiology at the Copenhagen Agricultural College. Intrigued with Francis Galton's "Theory of Heredity," Johannsen's most influential research was concerned with the effects of selection on self-fertilizing systems. His research on the Princess bean, *Phaseolus vulgaris*, demonstrated that progeny from a "pure line" were genetically identical; any differences were due only to environmental effects and not inheritable. He found that in the pure lines, selection was an ineffective force in producing new forms, since there was not enough variability to work upon. James A. Peters wrote of Uber Erblichkeit in populationen und in reinen Linien: "This thorough and meticulous investigation of the true significance of selection was a bombshell to evolutionary thought." He further wrote, "We owe to Johannsen our modern viewpoint of selection as a primarily passive process, which eliminates but does not produce variations."[2]

In 1905, Johannsen was appointed professor of plant physiology at the University of Copenhagen, and was made rector of the university in 1917, both these positions despite the fact that most of his knowledge was self-taught. *"Elemente der exakten Erblichkeitslehre"* was published in 1909. In it, Johannsen introduced the terms gene, genotype (to designate the genetic makeup of an individual), and phenotype (to describe the physical appearance of an individual, the interaction between the genotype and environment). A revised and expanded edition of some of his other work, *Arvileghedslaerens elementer* was the most influential European textbook of genetics of its time (1909). It was revised through several editions, the third and last appearing in 1926.

Johannsen did no more experimental work after his pure lines research. He continued to write and was a recognized critic and historian of science. He received several honorary degrees and was a member of the Royal Danish Academy of Sciences. He died in Copenhagen November 11, 1927.

Selected Works by Wilhelm Johannsen

(Note: Most of the biographical material about Johannsen is in Danish, and most of his work was published in either Danish or German.)

Om arvileghed og variabilitet. (On Heredity and Variation.) Copenhagen, 1896. Enlarged as *Arvileghedslaerens elementer*. Copenhagen, 1905. Rewritten and enlarged (in German) as *Elemente der exakten Erblichkeitslehre*. Jena, 1909. 3rd ed., 1926.

Uber Erblichkeit in Populationen und in reinen Linien. (Jena, 1903.) Published in abridged English translation in: James A. Peters, ed. *Classic Papers in Genetics*. Englewood Cliffs, N.J.: Prentice-Hall, Inc., 1959. pp. 20-26.

Selected Works About Johannsen

Dunn, L.C., "Wilhelm Ludwig Johannsen." In *Dictionary of Scientific Biography*. v.7: 113-115. New York: Charles Scribner's Sons, 1973.

> Concise biographical essay that presents Johannsen's diverse interests and talents: theorist, analytical logician, discoverer of "pure line" theory in genetics, and later as critic, clarifier, popularizer, and historian of science. Attributes his importance to genetics to his ability to grasp what would later become essential concepts: "Through the development of population genetics, the genetic structure of populations, as Johannsen was one of the first to recognize, occupies an increasingly important place in evolutionary biology."

Winge, Ojvind, "Wilhelm Johannsen." *Journal of Heredity* 49: 83-88 (1958).

> Personalized account of the man and his work by a friendly colleague.

BARBARA MCCLINTOCK

Barbara McClintock, American geneticist, was born on June 16, 1902. Her father was a physician and her mother taught the piano. Though her parents initially disapproved of her wish to attend college, they had a great respect for her independence and self-determination, and in 1919, Barbara enrolled in the College of Agriculture at Cornell University. She continued graduate work in the Department of Botany (Plant Breeding, where genetics was taught, did not at the time take women) majoring in cytology and minoring in genetics and zoology. While working as an assistant to a cytologist, she discovered a method for distinguishing maize chromosomes, and her future path was set. In 1927 she received her PhD and was offered an instructorship at Cornell.

McClintock's ambition was to synthesize research in the areas of cytological observation of the chromosomes with the observed inheritance patterns of the plants. Though plant breeders and cytologists were pursuing each aspect of the problem, they were doing so separately. Parallel work on *Drosophila* was already well-advanced. From 1929-1931 she published nine papers correlating observed genetic traits with morphologic markers on the maize chromosomes. In 1931, at the urging of T.H. Morgan of Columbia, a famous geneticist, she and her graduate student Harriet Creighton published "A Correlation of Cytological and Genetical Crossing-over in *Zea mays*." James A. Peters called this paper "A cornerstone in experimental genetics."[3]

A two-year fellowship from the National Research Council in 1931 followed, allowing her to study at the University of Missouri, at the California Institute of Technology, and again at Cornell. In 1933 she received a Guggenheim award and traveled to Berlin to work with Richard B. Goldschmidt, the controversial director of the Kaiser Wilhelm Institute. The trip was unsuccessful: McClintock was badly shaken by her observations of Nazi Germany, and she returned home to Cornell before the scheduled time elapsed. A Rockefeller Foundation grant allowed her to continue research at Cornell until 1936. In 1936, McClintock received a faculty appointment at the University of Missouri, where she stayed until 1941.

She was then offered a research position at the Carnegie Institute of Washington at Cold Spring Harbor on Long Island and has lived and worked there since. In the forties she began to receive recognition for her work: she was elected the first woman president of the Genetics Society of America in 1944 and in 1945 was only the third woman elected to the National Academy of Sciences. At Cold Spring Harbor, McClintock did the research leading to her discovery of movable genetic elements. Her research was ahead of its time however, and when she presented it in the 1950's there were few of her colleagues who understood and accepted it. A dozen years later, instances of transposable genetic elements were discovered in bacteria. Findings corroborating her observations continued to mount in the 70's and 80's: transposable elements or "jumping genes" were found not only in bacteria, but in yeast, *Drosophila*, and plants.

Thirty years after her first presentations of the material, Barbara McClintock was awarded the Nobel Prize in Medicine or Physiology in 1983 for her work on transposable genetic elements. McClintock is widely acknowledged as a brilliant geneticist whose work was decades before it was popularly acceptable. Other awards she has received include the Kimber Genetics Award from the National Academy of Sciences in 1967, the National Medal of Science in 1970, the Albert Lasker Medical Research Award in 1970, Israel's Wolf Foundation Prize in Medicine in 1981, and the Louisa Gross Horowitz Prize of Columbia in 1982. She also received in 1981, a lifetime, tax-free annual fellowship of $60,000 from the MacArthur Foundation.

The implications of McClintock's research are enormous: transposable elements may be involved in genetic regulation and might account for rapid speciation in plants and animals.

Selected Works by Barbara McClintock:

"A Cytological and Genetical Study of Triploid Maize." *Genetics* 14: 180-222 (1929).

"A Correlation of Cytological and Genetical Crossing-Over in *Zea mays*." Harriet Creighton and Barbara McClintock. *Proceedings of the National Academy of Sciences* 17: 492-497 (1931).

"The Origin and Behavior of Mutable Loci in Maize." *Proceedings of the National Academy of Sciences* 36: 344-355 (1950).

"Some Parallels between Gene Control Systems in Maize and in Bacteria." *American Naturalist* 95: 265-277 (1961).

Selected Works About McClintock:

Keller, Evelyn Fox. *A Feeling for the Organism*. San Francisco: W.H. Freeman and Co., 1983.

> This biography is a thoughtful, engaging account of McClintock's activities in, and conceptions of science. Chapters providing a descriptive context of genetic research at the time are interspersed with those chronicling the development of her research, allowing the reader to grasp the pattern of McClintock's work against the fabric of the time. Difficulties and disappointments McClintock experienced in a male-dominated field are depicted.
>
> The portrait that emerges is that of an original thinker, stubbornly self-reliant and independent to the point of continuing research on her self-appointed trajectory even after it was repeatedly ignored. This account can be enjoyed and understood by people with various backgrounds, as a technical grasp of biology is not assumed. The book has been translated into Spanish and French.

Lewin, Roger. "A Naturalist of the Genome." *Science* 222: 402-405 (October 28 1983).

> Following her Nobel Prize award, this review of McClintock's work is based heavily on her later discoveries of transposable genetic elements. It includes favorable commentary on her work by colleagues, particularly Marcus Rhoades, who worked in the maize cytogenetics group at Cornell in the late 1920's. The tone is not completely technical, though readers with some background in genetics will find this easier to read than will others. McClintock's originality and persistence are stressed, as is the progressive nature of her research.

N. I. VAVILOV

Nicolai Ivanovich Vavilov was born on November 26, 1887 in Moscow. Though seven children were born into the family, only four survived, and all grew up to be scientists.

Vavilov attended Petrovsky Agricultural Institute in Moscow, where he worked with a professor who was striving to identify more productive varieties of wheat, barley, and oats. In 1912, Vavilov began an official trip abroad, studying at London with William Bateson, one of the most well-known geneticists in the world at that time. While in London, Vavilov finished his post-graduate thesis on immunity in wheat. In 1917 he took a teaching post at Saratov, the capital of Russia's wheat-producing region. Though the area was beset with difficulties during the civil war, Vavilov was a popular teacher and was soon surrounded with colleagues and students. During this time he worked on his hypothesis that analogous variations occur in related species. This work was published as *The Law of Homologous Series in Variation* in 1922.

In 1921 Vavilov went to Petrograd (now Leningrad) to direct the Bureau of Applied Botany. By 1922, heading a famine-struck country, Lenin had decided that Soviet Russia needed a major agricultural research center. In 1925, after his death, the All-Union Institute of Applied Botany was set up, with Vavilov serving as its first director. From 1924 to 1929 Vavilov led a number of plant gathering expeditions to various parts of the world, including Afghanistan, Algeria, Tunisia, Morocco, Spain, Portugal, the south of France, Italy, Greece, Syria, Jordan, Ethiopia, China, Japan, and Korea. Vavilov suggested that there are concentrated centers of genetic diversity and that these places were the regions in which agriculture first began. He proposed that in all, there were 12 of these "centers of origin." Vavilov's expeditions netted thousands of varieties of wild plants and specimens of wheat. He became the first president of the Soviet Academy of Agricultural Sciences and was internationally renowned for his attempts at acquiring and conserving genetic variability for the purposes of crop improvement.

Although Vavilov was a respected scientist internationally and in Russia, his work was quenched by the infernal intrusion of politics into biology. In the late 1930's Trofim Lysenko, a plant breeder of humble birth and insubstantial scientific background became the spearhead of the "progressive biology" movement which was gaining momentum in Soviet Russia. Lysenko was backed by a philosophical campaign which chose Lamarck's ideas of inheritance of acquired characteristics as a suitable platform. Although Darwin's ideas of evolution were acceptable in Marxist thought, the ideas of Darwin and Lamarck did not mesh well so Lysenko and his retinue adapted Lamarckian modes to support socialist reconstruction. Though established scientists were initially displeased with Lysenko's disregard of genetic theory, his campaign prevailed precisely because it was custom made to impress Stalin.

In 1938, Lysenko was make President of the All-Union Academy of Agricultural Sciences. He and his supporters attacked Vavilov at every opportunity, deriding his scientific contributions and alleging that he was connected with enemies of the Soviet Union. From 1934 to 1940, at least eighteen of the staff scientists at Vavilov's Institute of Plant breeding were arrested. He was arrested in 1940 while on a plant collecting trip. Though originally sentenced to death, his sentence was commuted to detention in a labor camp for 20 years. He died in a prison camp on Jan. 26, 1943.

Selected Works by N.I. Vavilov:

"The Law of Homologous Series in Variation." *Journal of Genetics* 12, no. 1 (1922).

The Geographic Origins of Cultivated Plants. Leningrad, 1926.

"The Origin, Variation, Immunity and Breeding of Cultivated Plants." *Chronica botanica* 13, nos. 1-6 (1951). Published posthumously.

Selected Works About Vavilov:

Dobzhansky, Th. "I. Vavilov, A Martyr of Genetics: 1887-1942." *The Journal of Heredity* 38: 226-232 (1947).

   This is a short testament of Vavilov's great capacity for research and his talent for program administration. Includes a concise account of Lysenko's success which led to the downfall of Vavilov, and a short analysis of the crisis in Soviet genetic research due to the Lysenko episode. The article is very interesting both for its temporal (it was written only about ten years after Vavilov's disappearance from the international genetics scene) and personal (Dobzhansky was a Russian expatriate who was famous in evolutionary genetics) perspectives.

Popovsky, Mark. *The Vavilov Affair*. Hamden, Connecticut: Archon Books, 1984.

A lively and engaging account of the tragic consequences of political intrusion into scientific inquiry. Journalist Popovsky extracted and copied papers from a secret police file on Vavilov and conducted interviews with Vavilov's surviving friends and acquaintances to create this captivating book. Vavilov's early successes in research and in plant gathering expeditions are described, as his political decline from the presidency of the All-Union Academy of Agricultural Sciences. Much time is spent on the unraveling of Soviet thought in genetics as Trofim Lysenko climbed higher and higher in positions of scientific leadership. Scientific concepts are not treated in detail, this is more of an historical development than a scientific review.

Lysenko's rise to power is interpreted as a combination public relations campaign coordinated for him by Prezent, a social scientist with a background in philosophy, and an appeal to the approval of Stalin through formulating concepts of "progressive biology." This is a harrowing chronicle of what Andrei Sakharov called, "probably the ugliest episode in the history of science."

REFERENCES

1. *Man the Hunter*. R.B. Lee and I. De Vore, eds. Chicago: Aldine, 1968.
2. Peters, James A. *Classic Papers in Genetics*. Englewood Cliffs, N.J.: Prentice-Hall, 1959. p. 20.
3. Ibid., p. 156.

FURTHER READING

Allen, Garland E. *Life Science in the Twentieth Century*. New York: Wiley, 1975.

Carlson, Elof Axel. *The Gene: A Critical History*. Philadelphia: W.B. Saunders Company, 1966.

Mayr, Ernst. *The Growth of Biological Thought: Diversity, Evolution, and Inheritance*. Cambridge, Mass.: Belknap, 1982.

Harlan, Jack R. *Crops & Man*. Madison, Wisconsin: Crop Science Society of America, 1975.

Founders of Medical Techniques and Inventions: An Annotated Biobibliography

Andrea R. Testi

In compiling this biographical essay, an exemplary group of individuals has been chosen to epitomize founders of historic medical techniques and inventions. Scientists have been selected for their originality, tenacity, and poignant dedication to scientific methodology and experimentation. The majority have been honored with one of the highest accolades in science, the Nobel Prize Award. This article reflects upon the legacy of ten scientists who are recognized for exceptional achievements that conferred great benefits on mankind and substantially affected the course of medical research.

BEHRING, EMIL ADOLF VON
    b. March 15, 1854; Hansdorf, West Prussia
    d. March 31, 1917; Marburgh, Germany

Emil von Behring attended the Friedrich Wilhelms Institute, Berlin, where, in 1878, he received his M.D. degree. Directly entering the Army Medical Corps, he aspired to a lecturer position in the Army Medical College, Berlin in 1888. Within a year, he joined his compatriot, Robert Koch, at the Robert Koch Institute of Hygiene in Berlin. Between 1889 and 1895, Behring's theories on

Andrea Testi, MLS, is Collection Development Coordinator for Science and Technology in the Centennial Science & Engineering Library, University of New Mexico, Albuquerque, NM 87131-1466. She holds a Bachelor's degree in Psychology and a Master's degree in Library and Information Science from S.U.N.Y. Albany.

serum therapy and antitoxins were developed. In 1895 Behring was appointed Chair of Hygiene at the University of Marburgh, and Director of the Institute of Hygiene, Berlin. He was awarded the first Nobel Prize in Physiology or Medicine in 1901. Von Behring being responsible for establishing serotherapy as a medical specialty, has enabled researchers with endless paths for immunological exploration.

The Nobel Lectures in Immunology. The Nobel Prize for Physiology or Medicine, 1901, Awarded to Emil von Behring. *Scandinavian Journal of Immunology* 1989; 30(1):1-12.

In building a biographical collection highlighting the science of immunology, works of Emil von Behring are essential. This article is part of a series emphasizing researchers whose collective works are responsible for the "foundations on which the present-day science of immunology rests." Each article contains an introductory statement presented by a member of the Karolinska Institute, the original Nobel lecture speech, and a biographical paragraph exemplifying the award recipient. Appropriately so, the first lecture published is by Emil von Behring, the first Nobel Prize award in 1901 for Physiology or Medicine. This article allows the reader to travel back in time, reliving one of the most prestigious events in Behring's career.

Emil von Behring Born March 15, 1854. Arthur S. MacNalty. British *Medical Journal* Jan-June 1954; 1:668-670.

This article was written in celebration of the centennial of Emil von Behring's birth. Behring's pioneering efforts in immunization and his discovery of diphtheria antitoxin and toxin-antitoxin are outlined in a chronological format. The author effectively conveys the significant humanitarian benefits produced by Behring's discoveries.

BERGER, HANS

b. May 21, 1873; Neuses, now in Germany
d. June 1, 1941; Bad Blankenburg, now in East Germany

Hans Berger entered the University of Jena, Germany in 1892. He initially studied physics and astronomy, but later switched to the study of medicine; specializing in psychiatry. Upon graduation in 1897, Berger joined the University Psychiatric Clinic, Germany where he remained until his retirement in 1938. Starting as an assistant, Berger aspired to the ranks of Professor of Psychiatry and Director of the Clinic. His research career was guided by his continuing efforts to establish an association between the physical functions of the brain (e.g., blood flow and temperature) and the more subjective mental states of the brain. After much experimentation, in 1924, he discovered the human electroencephalogram (EEG). Berger's EEG discovery is a solid foundation for current neurological research, and is used routinely in diagnostic and therapeutic procedures.

Three Years with Hans Berger: A Contribution to His Biography. Raphael Ginzberg. *Journal of the History of Medicine and Allied Sciences* 1949; 4:361-371.

The author, Raphael Ginzberg first met Hans Berger in 1926. Berger was a full professor of psychiatry at the University of Jena, Germany. Ginzberg was resuming his medical studies at the University and attended Berger's psychiatric lectures. In March 1929, Ginzberg became a physician under the direction of Berger at the Jena Psychiatric Hospital, Germany. Based on his personal experiences working with Hans Berger and from staff feedback at the clinic, the author depicts Berger as "reticent and reserved; administering the hospital with rigid organization and discipline." However, in the evening Berger retreated to his private quarters, allowing his creativity to unfold. It was during these periods his scientific endeavors were achieved. Berger, an introspective man, kept his personal life and research experiments private. Hence, there is little in depth biographical information available. The author, based on his exposure, succinctly details Berger's character as manifested in his daily routines at the hospital.

Gloor, Pierre (Ed.). 1969. *Hans Berger on the Electroencephalogram of Man. The Fourteen Original Reports on the Human Electroencephalogram*. New York: Elsevier. 350 p.

The "father of electroencephalography," Hans Berger, discovered the electroencephalogram (EEG) in 1924. In 1929 he was the first to describe, in report format, the human EEG. Between 1929 and 1938, he generated 14 reports detailing his studies of the EEG. This text is a translation of these 14 original works; enriched with bibliographies, illustrations, and a detailed index. While not a biographical work in the purest sense, the author, rather than presenting the life history of Berger, presents a history of Berger's significant works on electroencephalography and behavioral science.

CARREL, ALEXIS

b. June 28, 1873; Lyons, France
d. November 5, 1944; Paris, France

Alexis Carrel earned his doctor of medicine degree in 1900 from the University of Lyons, France. A skillful surgeon, but lacking an interest in routine surgery, Carrel visited Canada in 1904 with intentions of becoming a cattle rancher. However, later that year he moved to Chicago where, at the University of Chicago, he began aggressive surgical experimentation. Dr. Carrel was noted for his technical originality, specifically for developing surgical suturing techniques for arteries and veins. He received an appointment at the Rockefeller Institute for Medical Research, New York in 1906. Carrel, often referred to as the "pioneer of vascular surgery" was awarded the Nobel Prize for Physiology or Medicine in 1912. This was the first Nobel Prize in medicine awarded to the United States. Carrel's suture techniques advanced vascular surgery and proved essential for future transplant surgery. During World War I, Carrel achieved another major research accomplishment. While serving as a surgeon in the French Army, Carrel and chemist Henry Dakin, developed the "Carrel-Dakin Antiseptic" for the treatment of deep wounds. In 1935, along with Charles Lindbergh, the aviator, he devised a "perfusion pump," the predecessor of the current mechanical heart. At the time of his death he was the Director of the Vichy Government's Carrel Foundation for the Study of Human Problems in Paris.

W. Sterling Edwards. 1974. *Alexis Carrel Visionary Surgeon.* Springfield, IL: Charles C Thomas. 143 p.

W. Sterling Edwards provides the reader with a thoughtful and comprehensive text, emphasizing the enthusiasm, interests, and accomplishments which formed the character of Alexis Carrel. In compiling this biography, Edwards selected information from original French biographies of Carrel, from the "Carrel Collection" at Georgetown University, Washington, D.C., and from personal interviews of Carrel's friends and colleagues. The impetus for this book was to fill a gap in the literature; to publish the first complete English language monographic biography of Alexis Carrel. Edwards, successful in accomplishing this goal, has generated a colorful essay that will attract all levels of readers.

Alexis Carrel: A Cardiovascular Prophet Crying in the Wilderness of Early Twentieth Century Surgery. Lyman A. Brewer. *Journal of Thoracic and Cardiovascular Surgery* 1987; 94:724-26.

This historical vignette focuses on Alexis Carrel, a "visionary pioneer vascular surgeon." By outlining in tabular form the dramatic accomplishments of Carrel's career, the author captures his limitless imagination and foresight in surgery. The belief that Carrel's techniques were highly advanced for his time, out of the grasp of most of his peers, is a prominent theme of this article.

Theodore I. Malinin. 1979. *Surgery and Life: The Extraordinary Career of Alexis Carrel*. New York: Harcourt. 242 p.

The reader is exposed to a fascinating view detailing Alexis Carrel's life and scientific accomplishments. While achieving extraordinary advances in all levels of his research, Carrel was surrounded by criticism and jealousy. The author, exemplified in his heartfelt writing style, shows his dedication to preserving Carrel's memory. This text is an asset to the literature on Alexis Carrel and surgical heritage.

EHRLICH, PAUL

  b. March 14, 1854; Strehlen, Upper Silesia
  d. August 20, 1915; Bad Homburg, Germany

Paul Ehrlich, after having attended four medical schools, received his medical degree from the University of Leipzig. Ehrlich's contributions to medical research are so vast that they are commonly divided into three groups: (1) the application of staining to the differentiation of cells and tissues for the purpose of revealing their function (1877-1980); (2) immunity studies (1890-1900); (3) chemotherapeutic devices (1907-1915) (W. Bulloch 1938). Based on his extensive research, Ehrlich is known as the founder of modern hematology. He is also known as the founder of chemotherapy by merit of his discovery of Salvarsan (606), an early cure for syphilis. Ehrlich is also highly regarded for his contributions to immunology and was jointly awarded the Nobel Prize for Physiology or Medicine in 1908.

Martha Marquardt. 1951. *Paul Ehrlich*. New York: Henry Schuman. 255 p.

The author, Martha Marquardt was Ehrlich's private secretary for thirteen years. Her goal, to uniquely portray the "beautiful and simple human soul" of Ehrlich, was achieved through this tender reflection of his life. Based on her intimate experiences, and by reading Ehrlich's personal notes and unpublished papers, Marquardt provides a complete picture of Paul Ehrlich's passionate and curious character. A collection of family photos as well as reprints of notes in Ehrlich's pen, serve to further personalize this text.

Paul Ehrlich Centennial. Roy Waldo Miner (Ed.) *Annals of the New York Academy of Sciences* 1954; 59:141-276.

This series of papers was a result of a conference held in honor of the centennial of the birth of Paul Ehrlich by the Section of Biology of The New York Academy of Sciences. The aim of the conference, documented in this text, was to outline Paul Ehrlich's work and its influence on experimental medicine. The papers written by some of Ehrlich's coworkers address his unique achievements and the evolution of his scientific concepts. The articles are highly technical in nature, thus best suited for a specialized science audience.

Paul Ehrlich as a Commercial Scientist and Research Administrator. Jonathan Liebenau. *Medical History* 1990; 34:65-78.

Paul Ehrlich has been eulogized as a biochemist, immunologist and medical researcher. His ties with industry and commercial work, however, have been discussed to a much lesser extent. This article is unique in that it presents a biography of Ehrlich as a commercial scientist and goes further in narrating a view of "medical science as a corporate activity."

EINTHOVEN, WILLEM

b. May 21, 1860; Semarang, Java
d. September 28, 1927; Leiden, Netherlands

Willem Einthoven, affectionately referred to as the "father of electrocardiography," received his doctor of medicine degree in 1885 from the University of Utrecht. In 1886 he accepted the position of chair of physiology at the University of Leiden. Einthoven first constructed a string galvanometer in 1903; the essential part of the modern electrocardiograph (ECG). While he continued to refine this instrument for numerous years, it was used immediately for general use. In order to understand deviations from the normal heart tracing, Einthoven devoted considerable time in securing an accurate interpretation of the normal result. By comparing abnormal readings to the norm, Einthoven was able to use the ECG as a diagnostic tool. His persistence in research was recognized in 1924 with the Nobel Prize for Physiology or Medicine for the development of the mechanism of the electrocardiogram. Einthoven's initial discoveries continue to inspire modern electrocardiography today.

Willem Einthoven (1860-1927): Some Facets of His Life and Work. Carl J. Wiggins. *Circulation Research* 1961; 9(2):225-233.

This paper was read at the Sixth Interamerican Congress of Cardiology (1960), commemorating the centennial of Willem Einthoven's birth. Rather than providing a comprehensive discourse on all facets of Einthoven's life, the author has chosen to limit his scope to the following areas: (1) the impact of Einthoven's work on cardiac research and diagnosis, (2) Einthoven, the man, and (3) events that helped shape his career. The article, while geared toward a specific audience, has much broader implications and certainly can be of interest for a wide range of readers.

Einthoven Commemoration. A. de Waart. *American Heart Journal* 1961; 61(3):357-360.

In celebration of the centenary of Willem Einthoven's birth, the author, A. de Waart, presented this lecture at Leiden University. De Waart, a longtime friend and colleague of Einthoven, presents a copious collection representative of Einthoven's dignified life and grand ideals. The warmth and admiration the author held for Einthoven is clearly evident in this paper. This text was reprinted in Bethe's Handbuch der Physiologie.

Obituary. Willem Einthoven, M.D., Ph.D. Thomas Lewis. *The British Medical Journal* 1927; 2:664-665.

After Willem Einthoven informed the scientific community of his electrocardiographic research, scientists began explorations of Einthoven's work. In particular, Thomas Lewis actively began confirming and refining Einthoven's studies. These two men shared a strong friendship. When accepting the Nobel Prize, Einthoven said in regard to Lewis, "without his valuable contributions I doubt that I should have had the honor of appearing before you today." It was Einthoven's nature to give others the credit and honor they deserved. Lewis, as exemplified by the tenderness of the obituary, gave Einthoven in death the same respect Einthoven bestowed upon others in life.

GRAHAM, THOMAS

b. December 20, 1805; Glasgow, Scotland
d. September 11, 1869; London, England

Thomas Graham, the "father of colloid chemistry" was one of the most creative chemical thinkers of the 19th century. In spite of his father's desire for him to enter the clergy, Graham persisted in his studies of chemistry, receiving his M.A. in 1826 from Glasgow University, Scotland. He is most noted for his gas diffusion law published in 1829. Graham's pioneering works on phosphates and dialysis are exemplary as well; with his research on the later having profound effects on modern renal dialysis techniques. Graham was not a good public speaker and did little to promote his findings; thus his more gregarious contemporaries held the popular reputation of these findings. Regardless, his scientific methodologies and persistence were unmatched. Graham was well respected and sought after by peers and students alike. Many prominent scientists were guided by his encouragement and friendship.

Stanley, Michael. 1987. *The Making of a Chemist Thomas Graham in Scotland*. Blairgowrie, Scotland: Lochee Publications. 14 p.

This short biographical piece is a concise and accurate documentation of Thomas Graham's life. Basic scientific concepts rather than detailed experiments are accentuated. The text is easy-to read, and is appropriate for all readers.

R. Angus Smith. 1884. *The Life and Works of Thomas Graham*. Glasgow: John Smith. 114 p.

This text, a collection of 64 private letters written by Thomas Graham, is the closest work emulating a full life biography of the eminent chemist. The collection, owned by the author, contains intimate correspondences from Graham's early college days to his old age. This rare collection will preserve a special memory of Graham; one which the author held dear to his heart.

Swords, Michael Dennis. 1973. *The Chemical Philosophy of Thomas Graham*. Cleveland: Case Western University (Ph.D. Dissertation) 118 p.

In the introduction the author states the purpose of the thesis, "to present an overview of Graham's scientific career, to place his work in the context of the scientific ideas and research of his time, and to interpret his research direction and creative conclusions in terms of his values and beliefs." The author has successfully fused the scattered pieces of Graham's discoveries, providing a picture of his research as a whole. Thus, the reader can begin to understand the motivation behind Graham's philosophies and achievements.

HARVEY, WILLIAM

b. April 1, 1578; Folkestone, England
d. June 3, 1657; Roehampton, England

Harvey attended the University of Padua, Italy, the greatest medical school of that time, receiving his M.D. degree in 1602. In 1628 Harvey published his masterpiece, De Motu Cordi et Sanguinis in Animalibus (On the Motion of the Heart and Blood in Animals). This announced one of the most profound discoveries of the modern period in anatomy and physiology, the circulation of the blood. Harvey's expert experimental techniques coupled with his abilities to expound his ideas and discoveries in the tenure of current scientific opinion are notable characteristics that will serve to depict Harvey across all time.

Keynes, Geoffrey. 1966. *The Life of William Harvey*. Oxford: Clarendon Press. 483 p.

In 1897 an excellent short biography of Harvey was published by Sir D'Arcy Power. After that point, no serious efforts to reconstruct his life were made. However, in 1942, Keynes was fortunate to acquire many of the source books for Harvey's life when Power's library was sold upon his death. Keynes, by utilizing sources Powers compiled and then expanding on this, was able to finish the work on Harvey. This text does not elaborate in any depth on Harvey's scientific work as many volumes have already been dedicated to this. Rather, it is more of a discourse on the path his life took, his family influences, his education and his collegial relationships. The text recapitulates Harvey's life in a cursive nature, adaptable to a multifarious audience.

Bylebyl, Jerome J. (Ed.). 1979. *William Harvey and His Age: The Professional and Social Context of the Discovery of the Circulation*. Baltimore: Johns Hopkins University Press. 154 p.

This book was generated from three essays presented at the May 1978 Annual Meeting of the American Association for the History of Medicine. Each chapter, written by a different author, emphasizes a unique phase in Harvey's career. His strong inclination toward basic medical sciences research, its direct applications on his clinical practice, and the effects on the medical community are discussed. The editor has done a skillful job of weaving a common theme into this text; the strong interrelationship between Harvey's scientific exploits and his identity as a practicing physician.

Whitteridge, Gweneth. 1971. *William Harvey and the Circulation of the Blood*. New York: Elsevier. 270 p.

This book provides an encyclopedic rendition of the scientific environment which guided Harvey's experimental theories. The book finds its strength as a reference tool focusing on the general pattern of medical education in the 16th and 17th centuries and the structure of the medical atmosphere during the 17th century. The reader is provided with an understanding of Harvey's precise experimental techniques rather than an interpretation of his character.

KOCH, ROBERT
b. December 11, 1843; Klausthal, Hanover
d. May 27, 1910; Baden-Baden

Robert Koch received his medical degree from the University of Gottingen in 1866. He was appointed district medical officer in Wolstein, Prussia, where he began his concentrated research efforts. Koch, noted for his vastly improved techniques of staining, photographing and cultivation of bacteria, perfected the agar plate method. This constituted the core of Koch's technique, paving the way for some outstanding advances in the conquest of diseases caused by bacteria. In 1882 Koch discovered the bacillus responsible for tuberculosis. This work earned him the honor of receiving the Nobel Prize for Physiology or Medicine in 1905. Koch had demonstrated, for the first time, a strict relationship between a microorganism and a human disease. His stringent and logical principles establishing an organism as the etiology of disease, have been distinguished as "postulates." These postulates, established in 1890, originated clinical bacteriology as a quantitative scientific discipline. Current procedures for studying bacteria, are essentially unchanged from those techniques devised by Koch.

Brock, Thomas D. 1988. *Robert Koch: A Life in Medicine and Bacteriology*. Wisconsin: Science Tech Publishers. 364 p.

This text, spotlighting the founder of medical bacteriology, fills a gap previously evident in the literature. In researching the history of microbiology, the author, Thomas Brock, noted the lack of any English language biography on Robert Koch. Thus, the current work was begun. The reader is guided through Koch's career noting his practical contributions in hygiene and public health. The personal side of Koch, following his progression from an eager country doctor to an opinionated, crusty old man, is brought to light. The author's matter-of-fact writing style makes the text suitable for a general readership level. Excellent photographs from the Robert

Koch Archives and the final chapter assessing the implications of Koch's work are prominent features of this book.

Carter, K. Codell. 1987. *Essays of Robert Koch*. New York: Greenwood Press. 189 p.

This text is a translated collection of a select group of Robert Koch's postulates. The essays, chosen for their relatedness, provide an excellent sample of Koch's finest contributions and display the consistency of his research. While in theory, not a true biographical work, providing the reader with an understanding of Koch's scientific theories can only help to enhance his overall biographies. The content is highly technical, thus most appropriate for readers with a scientific background.

KOCHER, EMIL THEODOR

b. August 25, 1841; Bern, Switzerland
d. July 27, 1917; Bern, Switzerland

Emil Theodor Kocher was granted his medical degree from the University of Bern, Switzerland in 1865. While studying surgery, he was under the mentorship of some of the most notable surgeons of that time, in particular, Theodor Billroth. In 1872 he was awarded a professorship of clinical surgery at the University of Bern. In 1911 he retired from his professorship but remained head of the University Surgical Clinic till his death. His first contribution to surgery that summoned public attention was the method used for the reduction of a dislocated shoulder. This sparked his incredible career of devising new and modifying older methodologies in the surgical arena. Kocher's famous surgical work on the thyroid gland culminated with the award of a Nobel Prize for Physiology or Medicine in 1909. Emil Kocher is also credited for the invention of numerous instruments and appliances, one of which, the "Kocher forceps," remain in current use. Kocher's life was dedicated to scientific research and to the benefits bestowed on humanity.

Bonjour, Edgar. 1950. *Theodor Kocher*. Switzerland: Paul Haupt. 61 p.

The author, Edgar Bonjour, is a member of the Kocher family by marriage. This relationship provides the author with intimate details allowing him to emphasize the elegance of Kocher's distinctive character and presenting a lifelike portrayal of Kocher for the reader. The material is well-balanced and appropriate for all readers.

Theodor Kocher. E. Robert Wiese and Judson E. Gilbert. *Annals of Medical History* 1931; 3:521-529.

The authors were thorough by incorporating the diversity associated with Theodor Kocher's character. The article leads the reader through a labyrinth of representative samples, citing Kocher's ingenuity and breadth of knowledge. This skillfully woven biography provides an historical record, memorializing one of the greatest general surgeons of all times.

RONTGEN, WILHELM CONRAD

b. March 27, 1845; Lennep, Rhine Province, Germany
d. February 10, 1923; Munich, Germany

Wilhelm Conrad Rontgen received his Ph.D. in 1869 from the Polytechnic at Zurich, Switzerland. By 1900, he was the Chair of Physics and Director of the Physical Institute at Munich, Germany. In studying physics, Rontgen's research extended into many branches including elasticity, piezoelectricity, and capillarity. Rontgens most noted achievement was the discovery of the x-ray in 1895. He achieved international notoriety and was awarded the first Nobel Prize for Physics in 1901. The memory of Rontgen's good nature and far-reaching discoveries will live forever.

Glasser, Otto. 1934. *Wilhelm Conrad Rontgen and the Early History of the Roentgen Rays*. Illinois: Charles C Thomas. 494 p.

X-rays were discovered by Rontgen in November 1895. By early 1896, news of the discovery had reached international acclaim. This year saw an unparalleled enthusiasm among scientists and laypeople alike. Recognizing the revolutionary affect of this discovery, the author strives to capture the events fixated around Rontgen and the "rays" during 1896. An impressive total of 1044 monographs and journal articles were generated on the topic. Chapter XXII provides the reader with a bibliography of these works. This rapid dissemination of knowledge exemplifies the immediate effect this discovery had on the scientific community. Glasser uses original photographs, most of which were published here for the first time, to provide a lifelike picture of these events. Otto Glasser is an eminently qualified radiologist known for his classic biographical works of Wilhelm Conrad Rontgen.

Nitske, W. Robert. 1971. *The Life of Wilhelm Conrad Rontgen Discoverer of the X Ray*. Tucson: University of Arizona Press. 356 p.

This detailed work by Nitske presents an intensive amount of information about Rontgen's scientific career and personal life. At times overwhelming, the text is extremely comprehensive and informative. Notable features include an extensive bibliography and lifelike photographic illustrations presenting the reader with a pictorial synopsis of Rontgen's life.

SOURCES

Asimov, Isaac. 1976. *Asimov's Biographical Encyclopedia of Science and Technology*. New York: Avon Books. 805 p.

Bulloch, W. 1938. *The History of Bacteriology*. London: Oxford University Press. 422 p.

Concise Dictionary of Scientific Biography. 1981. New York: Charles Scribner. 773 p.

Daintith, John et al. (Eds.). 1981. *A Biographical Encyclopedia of Scientists*. New York: Facts On File. 2v. 935 p.

The McGraw-Hill Encyclopedia of World Biography: An International Reference Work. 1973. New York: McGraw-Hill. 12v. various paging.

Millar, David. 1989. *Chambers Concise Dictionary of Scientists*. Cambridge, England: Chambers. 461 p.

Sourkes, Theodore L. 1966. *Nobel Prize Winners in Medicine and Physiology 1901-1965*. New York: Schuman. 464 p.

Walton, John et al. (Eds.). 1986. *Oxford Companion to Medicine*. Oxford: Oxford University Press. 2v. various paging.

SCI-TECH COLLECTIONS

Tony Stankus, Editor

During the 1980s one of the most popular television shows was named after its lead character, "Quincy," a forensic scientist played by Jack Klugman. Viewing it together with my friends who generally worked in the computer business and always felt the print library was one step away from total obsolescence — we would note Quincy, or his trusted sidekick Sam, injecting some chemical isolated from a corpse into a Varian and coming up with a printout of squiggly lines. After comparing this tracing to others in some tome, one of the stars would announce "Strychnine" or "heroin" or "digitalis" and the mystery would either be solved or initiated, depending on that night's plot. A number of my guests would question whether such a procedure was, in fact, really possible. I would then happily expound on spectra and my collection of spectral atlases, eager to finally explain a tool I worked with for a living, that had not yet been totally automated. My friends were not entirely incorrect in their forecasts of computer comparison searching of internally stored reference spectra in some machines ten years later. But print sources are scarcely threatened, given the vast numbers of spectra published and the combination of initial intuition and technical insight that must be used in interpreting initial categories of compounds within which to begin a search. Few collections of print spectra, moreover, seem as good as that of the University of Illinois at Chicago, and fewer still seem to be as well explicated as these by authors Odegaard and Hurd. Perceptive librarians such as these are the ultimately indispensable technology.

Spectra:
A Bibliography of Sources

Gladys Odegaard
Julie M. Hurd

SUMMARY. Spectra are important sources of information about atoms and molecules. This paper surveys the various types of spectra and the nature of what each reveals about atomic or molecular structures or properties. The instrumentation employed is identified. A brief history of the development of spectroscopy is included as well as a bibliography of sources for additional reading on the topic.

The second part of this paper is an annotated bibliography of published sources for the major types of spectra. Scope and coverage of each source is noted as well as details on the nature of access to the spectral data.

PART I.
SPECTRA AND SPECTROSCOPY: AN OVERVIEW

Early History of Spectroscopy

The earliest discoveries in spectroscopy were reported in writings on optics, as in the works of Sir Isaac Newton, Joseph Fraunhofer, and Robert Gustav Kirchhoff. Newton, whose work on spectra was first published in 1672, had wide-ranging scientific interests and for

---

Gladys Odegaard is Assistant Science Librarian and Assistant Professor, Science Library, University of Illinois at Chicago, Chicago, IL 60680. She received the MA in Library Science from the Graduate Library School, University of Minnesota.

Julie M. Hurd is Science Librarian and Assistant Professor, Science Library, University of Illinois at Chicago, Chicago, IL 60680. She received the MA in Library Science from the Graduate Library School, University of Chicago and the PhD in Chemistry from the University of Chicago.

a number of years pursued a study of the fundamental nature of light. Newton invented a telescope and while working with lenses had produced a number of prisms. During experiments with a triangular prism which he had ground, he observed the dispersion of white light (from sunlight) into a multi-colored image — a spectrum.

Other early spectroscopic experiments also represented efforts to understand and analyze light, frequently sunlight, but also that from incandescent sources such as flames. Fraunhofer, while engaged in work with lenses in the early nineteenth century, documented the dark lines that now bear his name interrupting the continuous spectrum of solar light. He published a detailed list of these lines by frequency and concluded that these were characteristic of sunlight. Some forty years elapsed before a partial explanation for these lines was provided by Kirchhof and Robert Bunsen who demonstrated a connection between the Fraunhofer lines and the spectra of metals. The fuller explanation that followed built on their work and ultimately attributed the dark Fraunhofer lines to absorption of light by the gaseous elements in the sun's atmosphere. This work provided the foundation for spectral analysis of distant astronomical objects and for the spectrochemical analyses now employed in countless scientific experiments. This early optical spectroscopy also led to techniques that allowed the testing of physicists' theories of atomic and molecular structure and led to a fuller understanding of atomic and molecular forces and energies.

Spectra as Information Sources

Atomic and molecular spectra are important information sources not only for chemists and physicists but for many other scientists who make use of the applications of spectroscopy to identify the presence of particular chemical species, whether in data returned from far-ranging space probes or from analyses of materials found within the human body or in core drillings from deep within the planet. Science and technology libraries frequently are visited by patrons seeking the spectra of known compounds or hoping to find in the published sources a spectrum that matches one they have measured. This article will describe some of the types of information that can be derived from spectra and will present an annotated bibliography of some of the major sources of spectra. While a science information specialist can find it helpful to have studied spectroscopy, such a technical background is not essential in helping patrons to locate spectra they are seeking. The introductory remarks which follow here should provide a context in which the non-chemist can interpret queries about spectra and make most efficient use of the many compilations and indexes that provide spectral information.

What Are Spectra?

In the earliest years of its history the term "spectrum" referred to the range of wavelengths perceived as colors of light by the human eye. As techniques and instrumentation developed, usage broadened to signify the entire range of electromagnetic radiation. In fact, the field now encompasses mass spectrometry as well where, rather than radiation emitted or absorbed, the instrumentation measures the distribution of charged particles produced after ionization of the sample.

Spectroscopy is an analytical technique that measures the interaction of energy with matter. As it is currently practiced, spectroscopy utilizes complex instrumentation to make and record these measurements — *spectra* — which are then interpreted to provide practical information such as definitive identification of the presence of specific atoms or molecules or to serve as the basis for theoretical discussions of the fundamental physical structure of chemical species. The practical applications of spectroscopy are widespread and used in many branches of pure and applied science and engineering. An understanding of the theoretical aspects of spectroscopy makes use of quantum mechanics and is a highly mathematical area of study pursued by physical chemists and chemical physicists. Many users of spectral data do so without this depth of understanding and, at some levels of application, are not handicapped by this lack.

A display of the data collected from a spectroscopic measurement is called a *spectrum*. A spectrum may be represented as a plot of intensity of radiant energy emitted or absorbed (or some function of intensity) versus the energy of that radiation (typically displayed as wavelength or frequency). Alternatively, a spectrum may be recorded as a series of numbers (representing emission or absorption peaks) that convey comparable information. Each chemical entity possesses its own unique spectrum. A spectrum may be thought of as the "fingerprint" of a species and it is frequently used in chemical analysis work as the basis for identification just as fingerprints are used to identify persons. Spectra can convey both qualitative information, i.e., the presence of a species, or quantitative information, i.e., the relative amounts of components in a mixture.

Types of Spectra

Spectra are classified (and sometimes named) according to type of radiation emitted or absorbed. Electromagnetic radiation is characterized by its wavelength which can vary from extremely long but low energy, as in the case of radio or microwaves, through the infrared and visible regions of the spectrum to very short wavelengths at the high-energy end of the spectrum as in X-rays or gamma radiation. (The energy of the radiation is inversely proportional to the wavelength.) Table 1 displays the principal regions of the electromagnetic spectrum and shows the units most conveniently used to measure wavelengths in each region. Each type of spectrum provides information on atomic or molecular energy levels, the distribution of species among energy levels, molecular geometries, chemical bonding, or interactions among molecules in mixtures or solutions. See Table 2 for the principal spectral regions, the instrumentation commonly employed in each, and the types of atomic or molecular transitions occurring in each region.

In the radio frequency and microwave regions of the spectrum the transitions observed relate to changes in either molecular rotational energy levels, nuclear spin levels (nuclear magnetic resonance or NMR), or electron spin levels (electron spin resonance or ESR). Both NMR and ESR spectroscopy measure the effects of magnetic fields on a sample. *Electron spin resonance spectroscopy* (also known as *electron paramagnetic resonance spectroscopy* or EPR) is concerned with the absorption of radiation by electrons in a molecule. ESR spectra can provide information on electronic structure and chemical reaction kinetics for paramagnetic species[1] including

FIGURE 1
The Electromagnetic Spectrum[1]

Spectral Region	Approximate Wavelength Range
Radio-frequency	10^{-1} - 10^3 meters (m)
Microwave	0.1 - 30 centimeters[2] (cm)
Infrared	2.5 - 50 micrometers[3] (μm)
Visible	400 - 800 nanometers[4] (nm)
Ultraviolet	200 - 400 nm
X-ray	0.05 - 1 nm
Gamma	< 0.05 nm

[1] McGraw-Hill Encyclopedia of Science and Technology. 6th edition. New York: McGraw-Hill; 1987. See the article on "spectroscopy," volume 17, pages 213-219.

[2] 1 cm = 10^{-2} m

[3] 1 μm = 10^{-6} m

[4] 1 nm = 10^{-9} m = 10 angstroms (Å)

FIGURE 2

Principal Types of Spectra[1]

Spectral Region	Radiation Source	Detector	Transitions
Radio-frequency	Radio transmitter	Radio receiver	Molecular rotations, nuclear magnetic resonance (NMR)
Microwave	Klystron	Silicon-tungsten crystal	Molecular rotations
	Magnetron	Bolometer	Electron Spin Resonance (ESR)
Infrared	Nernst glower	Thermocouple	Molecular vibrations
	Globar lamp	Bolometer	
Visible	Tungsten lamp	Photocell	Electronic excitation (atomic)
Ultraviolet	Hydrogen-discharge lamp	Photomultiplier	Electronic excitation (molecular)
X-ray	X-ray tube	Geiger counter	Ionization
Gamma	Radioactive nuclei	Geiger counter	Nuclear transitions and disintegrations
		Scintillation Counter	

[1] McGraw-Hill Encyclopedia of Science and Technology. 6th edition. New York: McGraw-Hill; 1987. See the article on "spectroscopy," volume 17, pages 213-219.

biological molecules, metals, complexes, and others whether in solid, liquid or gas phase. The technique is non-destructive and finds applications in solid state physics, organic and inorganic chemistry, materials science, biology, and environmental science.

Nuclear magnetic resonance spectroscopy originates from the effect of a magnetic field on atomic nucleii, most frequently either protons (i.e., ^1H) or ^{13}C. NMR spectroscopy has been used to elucidate structures of compounds important in numerous disciplines and has recently been developed as a non-invasive medical procedure that provides images of living organisms. This latter application is known as *magnetic resonance imaging* or MRI and is rapidly becoming a standard diagnostic device, especially in the area of cancer detection.

Microwave spectroscopy detects transitions between rotational energy states of molecules and, after analysis, offers data on molecular geometries, i.e., bond lengths and bond angles. The earliest work provided structural parameters for relatively low molecular weight organic species but the technique has more recently been extended to provide information on complex biologically-significant compounds as well as for free radicals and other unstable species.

Infrared spectra (IR) originate from vibrations (stretching, bending, rocking, etc.) of the bonds in molecules. Infrared spectroscopy provides information on functional groups present and their spatial arrangements that are of interest to scientists in many fields including chemistry, materials science, environmental science and engineering. Until the development of NMR spectroscopy in the 1950's, IR spectroscopy was the most important analytical technique for identification of species. It has seen a resurgence of interest with the development in the 1970's of *Fourier transform infrared spectroscopy* (FT-IR) and new commercially-available instrumentation that supports use of the technique in routine analyses. This type of spectrum, more than any other, is considered a molecular "fingerprint."

Raman spectra provide information on molecular vibrations and/or rotations complementary to that obtained from infrared spectra. Lasers are utilized as the source of radiation in the far infrared-visible region. This technique can be used to study metal bonding, ring compounds, steroids, and long chain molecules, among others.

Emission spectra observed in the *visible* region were among the earliest to be studied. They may be continuous spectra as emitted by incandescent solids, line spectra emitted by excited atoms, or band spectra emitted by excited molecules. The methods provide both quantitative and qualitative data for identification and analysis.

Ultraviolet (UV) and *visible* absorption spectra result from electronic transitions in molecules that produce broad bands best suited for qualitative study of simple solid or liquid samples whose molecules contain "chromophores"—structural fragments that absorb radiation at characteristic frequencies in this region. The spectrum obtained provides information on groups present or not present in the sample and is a function of the whole structure of the substance.

X-rays can cause substances to emit electrons and the energy distribution of these particles is a photoelectron spectrum which provides information on the electronic structure of the sample. In similar fashion *gamma rays* reveal details of nuclear structure.

Mass spectrometry is a destructive technique that separates gas phase ions produced from the sample being analyzed according to their masses or ratios of charge to mass. The resulting distribution of species resembles the spectra that are produced by emission or absorption of electromagnetic radiation and so the technique is frequently considered to be spectroscopic in nature; the terminology employed and the classification of library materials reflects that perspective. Mass spectrometry supports the identification of atomic and molecular chemical species even at the level of resolution that separates various isotopes of an element. This technique is another "fingerprint" procedure with many analytical applications. For these reasons sources of mass spectra are included in this bibliography.

Instrumentation

The types of instruments used to measure and record spectra are known variously as spectrometers, spectrophotometers, spectrographs, or spectroscopes and which instrument or term is employed in a given measurement depends on the spectral region. The process of recording a spectrum can be depicted schematically as shown in Figure 1. In any spectrometric experiment the sample (the species whose spectrum will be measured) must be properly prepared so as to interact with the radiation to which it will be subjected. This preparation may be simple, as in cases where the compound need only be placed in a suitable container (IR spectroscopy), or it may be more involved, even to the extent of decomposing a compound into its constituent atoms (mass spectrometry).

For absorption spectra, the prepared sample is subjected to electromagnetic radiation which is absorbed by the sample with a resultant increase in energy. Radiation not absorbed passes through the sample to a resolving device which separates the wavelengths present and allows a detector to analyze the intensity of radiation at each wavelength. The plot of intensity versus wavelength is an absorption spectrum and it displays lower intensities of transmitted radiation that correspond to the energy absorbed by the sample.

Emission spectra are obtained by employing an external energy source (heat, magnetic, mechanical, etc.) to raise the energy level of the sample. After the external source is removed, the sample then emits the energy gained as radiation: an emission spectrum. This emitted radiation passes through the resolving device and the measurement system. Recorded emission and absorption spectra, as noted above, are sources of both quantitative and qualitative information. The particular combination of wavelengths emitted or absorbed can reveal the presence of a specific chemical entity (qualitative); the intensity of the radiation provides information of a quantitative nature after suitable analysis.

As each of type of spectroscopy developed, the earliest instruments were built by those scientists who were the first to study that particular spectral region. As a specialization developed further, the instrumentation became commercially available, much in the same fashion as turn-key library automation systems followed individually-developed systems. Just as turn-key systems made it possible for more libraries to automate their activities, so commercially available spectroscopic equipment broadened the usage of spectroscopic techniques. Laboratories not directly involved in fundamental research in spectroscopy could purchase and use equipment that allowed increasingly sophisticated analyses of chemicals. These in-

FIGURE 3

Schematic Diagram of a Spectral Experiment

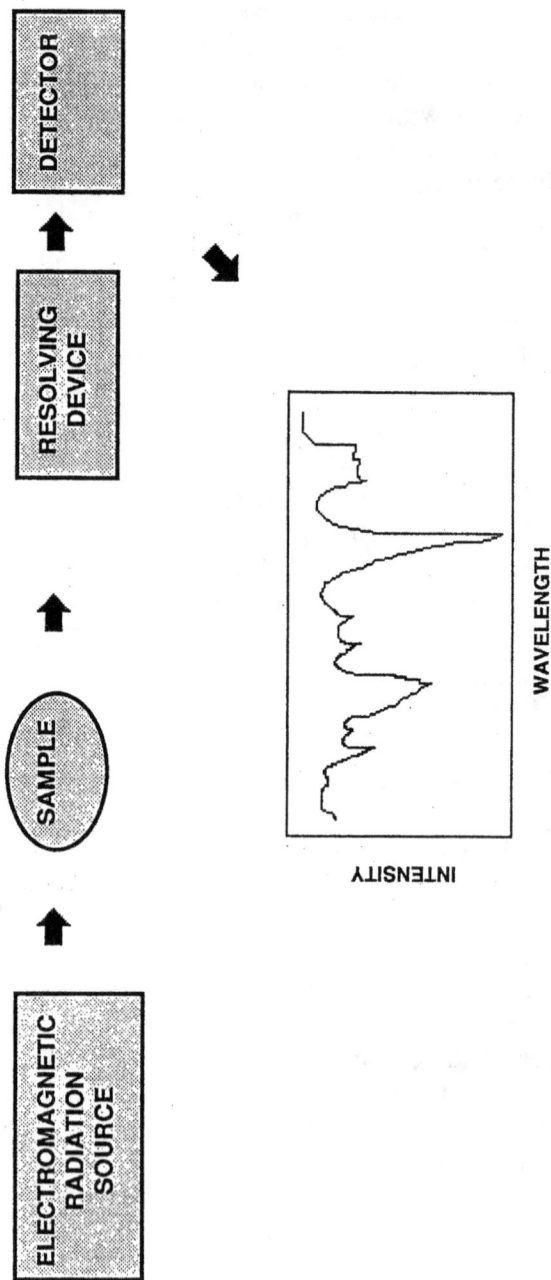

strumental techniques frequently replaced analytical techniques that utilized diagnostic reactions to identify chemicals present or quantitative methods employed to analyze multi-component mixtures. In turn, these new applications diversified the users of collections of spectra from the original chemists and physicists to include biologists, medical researchers, engineers and others. This fact should be recognized by information specialists working with materials on spectra because it may be important for a library to provide supplemental resources to assist some users of spectra collections. For example, a specialist in materials science may require guidance in identifying the systematic name for a chemical which is known in her laboratory only by its trade name.

Uses and Users of Spectra

Science library patrons seeking spectra or information about spectroscopy may range from the undergraduate chemistry student with a lab experiment in process to the research scientist seeking to identify a newly isolated compound. The library resources these patrons might consult could include introductory materials on spectroscopy to supplement a course text, organized compilations of published spectra, or the research literature in journals and technical reports. This article will suggest a few sources of introductory information to assist users in understanding spectra but will concentrate primarily on describing a selection of the numerous compilations of spectra. It must be noted here, however, that these are not exhaustive sources. As voluminous as several of the published collections are, not all compounds known are represented. There are many species whose spectra have not been measured or, if measured, may be found only in a journal article or in records of the scientist who determined the spectrum. Even the most skilled literature searcher may not be able to locate a sought-after spectrum; it may not be published and waiting to be found. It is the goal of this article to provide a list of sources sufficiently comprehensive that the search may be as exhaustive as available resources can support. After consulting without success the compilations listed in this article, a search of *Chemical Abstracts,* Beilstein[3] (for organic compounds), or Gmelin[4] (for inorganic compounds) may be required before it can be said with confidence that a published spectrum is not available. Searching these specialized reference sources will not be covered in this article as that is a topic deserving of fuller consideration and which is treated in a number of guides to the chemical literature.

Using Compilations of Spectra

There are certain features common to many of the compilations listed in the bibliography provided in Part II and these will be summarized here. As with any other reference work, these are best approached on first use by a careful examination of the introductory material. There one will usually find information on the scope and coverage of the work: what kind of spectra are included, for what type of compounds, taken from what sources, and arranged in what sort of order. The reader may also learn whether the compilation is an ongoing one that will be updated and, if so, with what regularity. It is essential to know what kind of access is provided to the spectra in a particular work. Some of the possibilities include: chemical name (common, trade, or systematic), chemical formula, molecular weight, major peaks or absorption lines, and Chemical Abstracts Service Registry Number. If a compilation is a multi-volume set, there may be cumulative indexes; if these do not exist it may be necessary to search individual volume indexes.

Information describing a compilation of spectra is not always easily found in its prefatory material. The extent to which an information specialist can learn about the features of a compilation influences the perceived utility of the reference in application to a specific problem. Put simply, if a scientist has only a spectrum and a chemical formula, those indexes with a chemical formula index and/or a listing of principle peaks are likely to offer a better beginning than those organized solely by chemical name. Conversely, the patron seeking a spectrum of a particular compound might best be referred to those works that provide chemical name or registry number indexes. In the bibliography that follows in the second part of this article the citations are annotated with information on scope, types of indexes, etc. to guide the users of those sources.

There are several publishers known for their very comprehensive compilations of spectra which are among those listed in the bibliography in part II. The Aldrich Chemical Company publishes spectral collections (*libraries*) that are organized by chemical categories based on chemical structures. Their libraries include NMR spectra, infrared spectra, Fourier Transform infrared spectra (FT-IR), and a carbon-13 NMR set. In addition, they offer a set of microfiche indexes to seven physical properties for Aldrich chemicals. All the compounds included in their publications are commercially available from the company. Indexes to spectral libraries typically include chemical name, molecular formula and CAS Registry Number. Full chemical names are provided for each species rather than trade names as are sometimes used in other sources. Because their inventory is extensive, the Aldrich collections can provide an easy-to-use first reference that offers a high probability of success for many chemicals.

The Sadtler Research Laboratories in Philadelphia have been publishing collections of spectra since 1947. Their comprehensive collections include over 342,000 spectra organized by type of spectra, UV, IR, NMR, ^{13}C NMR, Raman, etc., and the ongoing collections are issued in looseleaf binders of a particular shade of dark green well-known to users of their data. Their libraries are typically indexed by name, chemical formula, and chemical structure or group and use a Sadtler spectrum number as the basis for organization of the individual spectra. Sadtler also publishes a collective index to their entire set. In addition to the print format, machine-readable data for use with particular instruments or personal computers is also available. For those seeking a spectrum for a known compound, these comprehensive compilations promise a good chance of success.

Another publisher of specialized sources on spectra has been the American Petroleum Institute (API) which began in 1941, in cooperation with the National Bureau of Standards, to assemble data on physical, chemical, and spectral properties of hydrocarbons and other compounds of interest to the petroleum industry. The API Project 44 has grown over the years and is now located at the Thermodynamics Research Center (TRC) associated with Texas A & M University.

The sources of spectral information published by the above laboratories and others included in the bibliography that follows have appeared in book format, often in looseleaf binders that allow for additions of new spectra as these become available. Microform versions also have been available and are a more space-conserving storage medium. In recent years compilations of spectra have also begun to appear in machine-readable form supporting retrieval and manipulation of the data. These databases of spectra are likely to increase in number and size and very likely represent the future format for this type of information.

Books About Spectra

Listed here are books about spectra and spectroscopy that complement the sources of spectra themselves found in Part II of this paper. These materials range from very introductory level reference works that might be useful to undergraduate students to monographs used as course texts in graduate courses that teach about spectroscopy. It should be noted that only a very small selection of the latter type of work is included here; the titles listed have been chosen because they are among the more recent books published on the topic and may be less well-known than some of the classic works in the field. They serve as examples of a much larger body of literature. Additional introductory reading may also be found in any good scientific encyclopedia under entries such as "spectra," "spectroscopy," and the various types of spectra.

Acronyms and abbreviations in molecular spectroscopy. Detlef A.W. Wendisch. Berlin and New York: Springer-Verlag; 1990. 315 pages. Entries include definitions, applications, and literature references.

Analytical instrumentation handbook. Galen Wood Ewing, editor. New York and Basel: Marcel Dekker; 1990. 1071 pages. Chapters contributed by various analytical chemists describe instrumentation and techniques. Illustrations and bibliographies.

A Bibliography of matrix isolation spectroscopy: 1954-1985. David W. Ball et al. Houston, TX: Rice University Press; 1988. 643 pages. Bibliography of published references that deal with matrix-isolation spectroscopy with emphasis on cryogenic inert gas matrix isolation. Almost 3700 items in 3 section: Books, review articles, general references. Author, subject and formula indices.

The chemist's ready reference handbook. Gershon J. Shugar and John A. Dean, consulting editors. New York: McGraw-Hill; 1990. various paging. Describes instrumentation and analytical procedures in chapter arrangement with references and bibliographies.

[*CRC*] *Handbook of spectroscopy.* J.W. Robinson, editor. Boca Raton, FL: CRC Press; 1974. 2 volumes. Intended to provide a reference for the spectroscopic data available on the most important materials in the major fields of spectroscopy.

A dictionary of concepts in NMR. S.W. Homans. Oxford: Clarendon; 1989. 343 pages. "The aim . . . is to aid chemists and biochemists familiar with the basic principles . . . to understand the bewildering array of acronyms and technical jargon . . ." Biased towards two-dimensional NMR methods in liquids. Many entries include references for further reading.

A dictionary of spectroscopy. Ronald Denney. 2d edition. New York: Wiley; 1982. 205 pages. The most common expressions, equations and terms in an alphabetical format with literature references. Includes modern and traditional types of spectroscopy. Directed to students and those not generally familiar with developments in the field.

The encyclopedia of spectroscopy. George C. Clark, editor. New York: Reinhold; 1960. 787 pages. Topics arranged alphabetically according to principal kinds of spectroscopy and then by the various aspects. Signed articles.

Fundamentals of molecular spectroscopy. 2d edition. C.N. Banwell. New York and London: McGraw-Hill; 1972. 332 pages. Non-mathematical introductory treatment.

Fundamentals of molecular spectroscopy. Walter S. Struve. New York: Wiley; 1989. 379 pages. Introduction to spectroscopy of atoms and molecules for graduate students with some background in quantum mechanics and mathematics.

Group theory in spectroscopy: with applications to magnetic circular dichroism. Susan B. Piepho and Paul N. Schatz. New York:

Wiley; 1983. 634 pages. Advanced treatment of application of group theoretic methods to molecular spectroscopy.

A guide to collections and indexes of spectra. Irwin A. Rodin, compiler. Waterloo, ON: University of Waterloo. Engineering, Mathematics and Sciences Library; 1978. 25 pages. Annotated bibliography of collections of spectra and indexes to spectra in the EMS Library; will help locate spectra by name, formula, wavelength, shift, mass/charge or functional group.

Mass spectrometry: applications in science and engineering. Frederick A. White and George M. Wood. New York: Wiley; 1986. 773 pages. Interdisciplinary overview of modern mass spectrometry with historical background, instrumentation, and applications in engineering, physical sciences and life sciences.

Molecular spectroscopy. Jack D. Graybeal. New York: McGraw-Hill; 1988. 732 pages. Provides a foundation for beginning graduate students seeking a quantum mechanical treatment of molecular spectroscopy including rotational, vibrational, and electronic spectra as well as nuclear and electron resonance.

Molecules and radiation: and introduction to modern molecular spectroscopy. Jeffrey I. Steinfeld. Cambridge, MA and London: MIT Press; 1979. 348 pages. Graduate level introduction for chemical physics students to modern molecular spectroscopy intended to provide background for understanding of current research in the field.

Multilingual dictionary of important terms in molecular spectroscopy. International Union of Pure and Applied Chemistry. Commission on Molecular Structure and Spectroscopy. Ottawa: National research Council of Canada; 1966. 221 pages. English, French, German, Japanese and Russian.

Nineteenth Century Spectroscopy: Development of the Understanding of Spectra 1802-1897. William McGucken. Baltimore and London: Johns Hopkins,1969. 233 pages. Examines developments from the work of Wollaston with dark solar lines (1802) through discoveries by Busen and Kirchhof in 1860 up to J.J. Thomson's discovery of the electron in 1897.

Spectrometric identification of organic compounds. 4th edition. Robert Silverstein, G. Clayton Bassler and Terence C. Morrill. New York: Wiley; 1981. 442 pages. A how-to-do-it book. Identifications by means of mass, IR, NMR and UV spectroscopy.

Symmetry and spectroscopy: an introduction to vibrational and electronic spectroscopy. Daniel C. Harris and Michael D. Bertolucci. New York: Oxford; 1978. 550 pages. Introductory level text for upper level undergraduate and beginning graduate level chemistry students. Begins with chapter on group theory. Many examples and experimental results.

PART II.
BIBLIOGRAPHY

This bibliography lists collections of spectra and indexes to spectra grouped by type of spectra and organized within each group alphabetically by title. Sources of spectra in the visible region may be found in the sections on infrared and ultraviolet spectra, or in the section on "miscellaneous and various spectra," depending on how material was organized by source's compiler. This bibliography is based on the Spectra Collection maintained by the Science Library at the University of Illinois at Chicago and represents selected sources from that collection. The authors welcome suggestions of items that might be added to this compilation or information on new editions of sources included here.

Carbon-13 NMR Spectra

Atlas of carbon-13 NMR data. E. Breitmaier, G. Haas and W. Voelter. London, Philadelphia: Heyden; 1979. 3 volumes. Lists 1,003 organic compounds. Does not include spectra. Indexed alphabetically, by IUPAC name, chemical class, molecular formula, molecular weight, and chemical shift.

Carbon-13 NMR spectra; a collection of assigned, coded, and indexed spectra. Leroy F. Johnson and William C. Jankowski. Huntington, NY: R.E. Krieger; 1978. 580 pages. 500 spectra arranged in empirical formula order. Indexed by name, code (a coding system similar to that in the Varian proton NMR catalogs was used), and chemical shift.

Chemical shift ranges in carbon-13 NMR spectroscopy. Wolfgang Bremser, Burghard Franke and Hans Wagner. Weinheim, Deerfield Beach, FL: Verlag Chemie; 1982. 890 pages. Lists expectation ranges of chemical shifts and functions of partial structures; provides a means of predicting the positions of the C-13 NMR signals of almost any given structure. Printed version of the file from the COM-Microfiche collection Carbon-13 NMR spectral data published by Verlag Chemie. 23,450 compounds; about 240,000 shifts corresponding to 100,000 different substructure codes.

Sadtler guide to carbon-13 NMR spectra. William W. Simons, editor. Philadelphia: Sadtler Research Laboratories; 1983. 652 pages. A survey collection of 500 selected spectra with emphasis on chemical shifts. Ordering by functional groups, similar to other Sadtler handbooks. Alphabetical, numerical and molecular formula indexes.

Selected ^{13}C nuclear magnetic resonance spectral data. American Petroleum Institute. Research Project 44. College Station, TX: Thermodynamics Research Center, Texas A & M University; 1974-. Principally hydrocarbons and compounds of interest to the petroleum industry Indexed; also indexed by compound name and formula in the *Comprehensive index of API 44-TRC selected data.*

Standard carbon-13 NMR spectra. Philadelphia: Sadtler Research Laboratories; 1980-. 16 mm reel-to-reel microfilm or looseleaf. Spectra only. 19,000 spectra as of 1985. Indexed by *Standard carbon-13 NMR spectra collection indexes.*

Infrared Spectra

Absorption spectra in the infrared region. L. Lang, editor. London: Butterworths (volumes 1-2) and Huntington, NY: R.E. Krieger (volumes 3-4); 1974-78. 4 volumes.
Organic compounds of interest to the pharmaceutical and petrochemical industries. Volumes are individually indexed by substance, formula and author; each also includes list of literature references.

The Aldrich library of FT-IR spectra. Charles Pouchert. Milwaukee: Aldrich Chemical Company: 1985. 2 volumes.

Spectra for 11,000 compounds. Alphabetical, molecular formula and CAS registry number indexes.

Aldrich library of infra-red spectra. Charles Pouchert. 3rd edition. Milwaukee: Aldrich Chemical Company; 1981. 1873 pages.

Contains 12,000 spectra arranged by functional group in an order of increasing complexity.

Alphabetical and molecular formula indexes.

Atlas of infrared spectroscopy of clay minerals and their admixtures. H.W. van der Marel and H. Brudelspacher in collaboration with E. Reitz and P. Krohmer. New York: Elsevier; 1976. 396 pages.

Examples selected from about 4,000 samples totally investigated, most from well-known localities. Includes spectra. Extensive bibliography and subject index.

Atlas of polymer and plastics analysis. Dieter O. Hummel and Friederich K. Scholl. 2d, revised, edition Munich: Hanser and Weinheim: Verlag Chemie; 1978. 5 volumes.

V.1: Polymers: structures and spectra. — v.2: Plastics, fibres, resins: starting and auxiliary materials, degradation products. — v.3: Additives and processing aids. Spectra arranged by structure class with name and formula indexes. Bibliography.

ATR; attenuated total reflectance. Philadelphia, PA: Sadtler Research Laboratories; 1966-. Looseleaf.

1500 attenuated total reflectance spectra of commercially available monomers and polymers; volumes 1-2 are prism spectra, volumes 3-5 are grating spectra. Precise listing of chemical categories in introduction to index. Alphabetical index by trade name, and chemical name where applicable: chemical classification by type.

Evaluated infrared reference spectra (the Coblentz Society spectra). Joint Committee on Atomic and Molecular Physical Data. Printed and distributed by Sadtler Research Laboratories, Philadelphia, PA; 1961-. Looseleaf.

10,500 spectra. Indexed by spectrum number, name and molecular formula.

Handbook of infrared standards, with spectral maps and transition assignments between 3 and 2600 μm. Guy Guelachvili and K. Narahari Rao. Orlando, FL: Academic Press; 1986. 851 pages.

"It has long been recognized that one can make full use of the high resolution achieved in IR spectra only when there is corresponding precision available in the wavenumber standards used for calibrations. . . . This handbook provides lists of IR standards based on spectra of easily available molecular species . . ." — Introduction. Spectral maps, wavenumber tables, and transition assignments for CO, OCS, CO_2, N_2O, H_2O, $^{14}NH_3$, $^{15}NH_3$, and C_2H_4.

Index of vibrational spectra of inorganic and organometallic compounds. Norman N. Greenwood, E.J.F. Ross and B.P. Straughan. London: Butterworths: 1972. 3 volumes.

Approximately 25,000 compounds (IR and Raman). Spectra not included. Formula index.

Infrared absorption spectra; index, 1945-1957. H.M. Hershenson. New York: Academic Press; 1959. 111 pages.

Bibliographic references for spectra of 16,000 organic and inorganic compounds. Does not include spectra. Chemical substance index.

Infrared analysis of polymers, resins and additives: an atlas. Dieter O. Hummel and Friederich K. Scholl. New York: Wiley-Interscience; 1969. 2 volumes.

Part. 1: Text; Part. 2: Atlas. 1,774 compounds. Includes spectra. Chemical substance index.

Infrared band handbook. Herman Szymanski and Ronald E. Erickson. 2d edition, revised and enlarged. New York: IFI/Plenum; 1970. 2 volumes.

27,840 IR bands of organic and inorganic compounds. Does not include spectra. Formula index.

Infrared spectra and characteristic frequencies ~ 700-300 cm^{-1}. Freeman Bentley, Lee D. Smithson and Adele L. Rozek. New York: Interscience; 1968. 779 pages.

1,566 organic and inorganic compounds. Includes spectra. Formula index.

The infrared spectra atlas of monomers and polymers. Philadelphia, PA: Sadtler Research Laboratories; 1980. 810 pages.

Desk reference collection of 2000 IR reference spectra of monomers, polymers and precursors commonly encountered in industry and academia. Spectra are arranged into chemical class and in order of increasing complexity within each class. Alphabetical, numerical and Spec-Finder indexes.

The infrared spectra atlas of surface active agents. Philadelphia, PA: Sadtler Research Laboratories; 1982. 846 pages.

Reference collection of 2000 IR spectra of surface active agents encountered in both industry and academia. Substances identified by commercial names and chemical or descriptive names. Numerical, alphabetical and Spec-Finder indexes.

Infrared spectra handbook of common organic solvents. Philadelphia, PA: Sadtler Research Laboratories; 1983. 400 pages.

Infrared reference spectra of 400 common solvents grouped by chemical class, presented in a transmittance vs. frequency (wave number) format over the spectral region 4000-400 cm^{-1}. Alphabetical and molecular formula indexes.

The infrared spectra handbook of inorganic compounds. Philadelphia, PA: Sadtler Research Laboratories; 1984. 345 pages.

A desk reference collection of infrared spectra for 345 inorganic compounds, compiled by Sadtler Research Laboratories. Also includes pertinent chemical and physical data; alphabetical and molecular formula indexes.

The infrared spectra handbook of priority pollutants and toxic chemicals. Philadelphia, PA: Sadtler Research Laboratories; 1982. 727 pages.

564 IR reference spectra representing 363 chemical substances which have appeared in the EPA Priority Pollutants List, the OSHA Category I List of Carcinogenic Substances and a list of hazardous compounds are common to industry and of concern during interstate transportation. Some 200 of the compounds are represented by both condensed phase and vapor phase spectra. Compiled by Sadtler Research Laboratories. Alphabetical, molecular formula and Spec-Finder indexes.

Infrared spectra of inorganic compounds (3800-45 cm⁻¹). Richard Nyquist and Ronald O. Kagel. New York: Academic Press; 1971. 495 pages.

892 compounds. Includes spectra. Chemical substance index.

Infrared spectra of minerals and related inorganic compounds. J.A. Gadsden. London: Butterworths; 1975. 277 pages.

Contains IR band tables for over 600 minerals and 100 other related inorganic compounds, referenced to the literature where the original spectra can be found. Also includes references to a further 550 minerals and inorganic compounds. Subject/chemical and mineral indexes.

Infrared vapour spectra, group frequency correlations, sample handling and the examination of gas chromatographic fractions. David Welti. London: Heyden, in cooperation with Sadtler Research Laboratories: 1970. 211 pages.

Includes 306 vapor spectra in numerical serial order, with index to vapor spectra in Sadtler, DMS, Coblentz and Wyandotte systems.

Inorganics and related compounds. Philadelphia, PA: Sadtler Research Laboratories; 1967-. Looseleaf.

1300 spectra of organic and coordination compounds. Alphabetical and numerical indexes.

Interpretation of vapor-phase infrared spectra. R.A. Nyquist. Philadelphia, PA: Sadtler Research Laboratories; 1984. 2 volumes.

Intended to aid in the interpretation of vapor phase infrared spectra generated utilizing the gas chromatography/infrared technique. Vol. 1: Group frequency data; Vol. 2: Spectra (500 vapor phase IR reference spectra). Alphabetical and molecular formula indexes.

Lubricants; grating spectra. Philadelphia, PA: Sadtler Research Laboratories; 1966-. Looseleaf.

900 spectra. Alphabetical and class indexes.

Massachusetts Institute of Technology wavelength tables. Cambridge, MA: Massachusetts Institute of Technology. Spectroscopy Laboratory; 1939. 2 volumes.

Main tables contain more than 100,000 entries for the most important known spectrum lines emitted between 10,000 and 2000 angstroms by atoms in the first two stages of ionization. — Introduction. Emission lines measured by diffraction-grating spectrographs.

Massachusetts Institute of Technology wavelength tables. Volume 2: Wavelengths by element. Massachusetts Institute of Technology. Spectroscopy Laboratory. Cambridge, MA: MIT Press; 1982.
First stage of a project to expand and correct the 1939 edition. Presents the same atomic emission lines (as corrected in the 1969 edition) rearranged by element, with the addition of wavelengths in vacuum.

Molecular spectra of metallic oxides. Alois Gatterer, J. Junkes and E.W. Salpeter. Citta del Vaticano: Specola Vaticano; 1957. 91 pages.
Wavelength tables; portfolio of 36 plates. Principally spectra of astrophysical interest. Spectral range from 9000 to 2000 Å; grating spectrograph. Organized by compound. Analytical remarks rely heavily on Rosen's *Données spectroscopiques concernant les molecules diatomiques* published in 1951.

Pharmaceuticals grating spectra. Philadelphia, PA: Sadtler Research Laboratories; 1972-. Looseleaf.
1200 spectra. Indexed by name, molecular formula, and therapeutic application.

Standard grating spectra. Philadelphia, PA: Sadtler Research Laboratories; 1968-. Looseleaf.
79,000 spectra by 1990. Indexed through the *Sadtler cumulated comprehensive indexes.*

Sadtler handbook of infrared spectra. William W. Simons, editor. Philadelphia, PA: Sadtler Research Laboratories; 1978.
Abridged edition (approximately 3,000 spectra) of *Sadtler standard IR grating spectra.* 1035 pages.
Intended to serve as a small, convenient reference collection of IR spectra of organic compounds for college courses or when comprehensive collections are not available.

Sadtler infrared spectra handbook of esters. Philadelphia, PA: Sadtler Research Laboratories; 1982. 686 pages.
The term 'esters' as used to describe the contents of this volume

is applied only to the non-cyclic esters of carboxylic acids. Other types of esters will appear in future volumes dedicated to that group of compounds, e.g., sulfonic acid esters will be included with other compounds containing sulfur. 2,000 IR grating spectra selected from among those published in the *Standard IR grating collection* between 1966 and 1980. Alphabetical, Spec-Finder and numerical indexes.

Sadtler infrared spectra handbook of minerals and clays. John R. Ferrar, editor. Philadelphia, PA: Sadtler Research Laboratories; 1982. 440 pages.

A collection of 400 infrared spectra of various minerals taken in the region of 4000-250 cm^{-1}, determined using the KBr pellet technique. Presented in a linear transmittance vs. wave number format; classified according to an increasing order of complexity of the mineral. Indexed alphabetically by name of mineral.

Selected infrared spectral data. American Petroleum Institute. Research Project 44. Washington, DC: American Petroleum Institute, 194-. 2 volumes. Looseleaf.

Principally hydrocarbons and compounds of interest to the petroleum industry. 3,395 compounds through 1976. Indexed; also indexed by compound name and formula in the *Comprehensive index of API 44-TRC selected data.*

Tabulation of infrared spectral data. David Dolphin and Alexander Wick. New York: Wiley; 1977. 549 pages.

Changes in characteristic group frequencies. Does not include spectra, except for spectra of common solvents given in the last chapter. Wave number-micron conversion table on back fly-leaves.

Mass Spectra

Atlas of mass spectral data. E. Stenhagen, S. Abrahamson and F.W. McLafferty. New York: Interscience; 1969. 3 volumes.

Lists approx. 20,000 organic and inorganic compounds. Includes spectra. Formula index.

Compilation of mass spectral data. A. Cornu and R. Massot. London: Heyden in cooperation with Université de France; 1979. 2d edition. 2 volumes.

Lists 10,000 organic compounds. Does not include spectra. Formula index.

CRC handbook of mass spectra of drugs. Irving Sunshine, editor. Boca Raton, FL: CRC Press; 1981. 457 pages.

CI and EI data indexed alphabetically and by molecular weight; approx. 500 EI curves with an 8-peak index. Combined index to all tables by compound name.

Eight peak index of mass spectra; the eight most abundant ions in 66,720 mass spectra, indexed by molecular weight, elemental composition and most abundant ions. 3rd edition. Nottingham: the Mass Spectrometry Data Centre, the Royal Society of Chemistry; 1983. 3 volumes in 7.

Designed to find compounds that correspond to particular mass spectral data. Volume 1 (2 parts): Spectra in ascending molecular weight order, sub-ordered on number of carbon atoms, hydrogen atoms, etc. Volume 2 (2 parts): Spectra in ascending molecular weight order, sub-ordered on mass/charge values in order of decreasing abundance. Volume 3 (3 parts): Spectra in ascending mass/charge value order, where each mass/charge is the first and then the second most abundant, sub-ordered on the mass/charge values in order of decreasing abundance.

Handbook of static secondary ion mass spectrometry. D. Briggs, A. Brown and J.C. Vickerman. Chichester, NY: J. Wiley; 1989. 156 pages.

Section A discusses the principles of static SIMS. Section B provides reference spectra from a range of materials with attempted assignment of characteristic peaks. Section C provides more reference spectra and case studies of actual applications of SIMS.

Index and bibliography of mass spectrometry, 1963-1965. Fred W. McLafferty and John Pinzelik. New York: Interscience. 1 volume. Unpaged.

KWIC indexing.

Mass and abundance tables for use inn mass spectrometry. J.H. Benyon and S.E. Williams. Amsterdam: Elsevier; 1963. 570 pages.

Restricted to the four most common elements of organic chemistry (C,H,N,O); combinations restricted to those containing at least one atom of C and 6 or fewer of N or O, and H atoms not exceeding the number required to give a fully saturated molecule. All masses calculated relative to the mass standard $^{12}C = 12u$. Introduction in English, German, French and Russian.

Selected mass spectral data. American Petroleum Institute. Research Project 44. Washington, DC: American Petroleum Institute; 1947-. Looseleaf.

Principally hydrocarbons and compounds of interest to the petroleum industry. Standard and matrix formats. 2,259 standard spectra through 1975 and 432 matrix spectra through 1976. Indexed; also indexed by compound name and formula in the *Comprehensive index of API 44-TRC selected data.*

Tables for use in high resolution mass spectrometry. Robert Binks, J.S. Littler and R.L. Cleaver. London: Heyden in cooperation with Sadtler Research Laboratories: 1970. 160 pages.

Tables devised to help the operator of a high resolution mass spectrometer to find the exact masses and hence the elemental composition . . . of any peak in mass spectrum by comparing it with a peak of known composition in the spectrum of a reference compound. – Introduction. Inserted (in pocket): "Chemical formulae from mass determinations."

The Wiley/NBS registry of mass spectral data. Fred W. McLafferty and Douglas B. Slauffer. New York: Wiley; 1989. 7 volumes.

Combination of the revisions of two books and their data base versions: *Registry of mass spectral data* (Stenhagen, Abrahamson and McLafferty) and *EPA/NIH mass spectra data base and its two supplements* (Heller & Milne). "[an attempt] to obtain the mass spectral data of as wide a variety of compounds as possible . . . this collection has been limited to unit-resolution mass spectra obtained by ionization at conventional electron energies (~70eV)" – Preface. Arranged by nominal molecular weight and Hill/CAS system.

Indexed by compound name, molecular formula, and CAS registry number.

Nuclear Magnetic Resonance Spectra

Aldrich library of NMR spectra. 2d edition. Charles Pouchert. Milwaukee: Aldrich Chemical Company: 1983. 2 volumes.

Consolidates the eleven volumes of the first edition with an additional 2,000 spectra; therefore provides spectra for more than 4,000 compounds. Molecular formula, alphabetical and Aldrich catalog number indexes at end of volume 2.

Formula index to NMR literature. M. Gertrude Howell, Andrew S. Kende and John S. Webb, editors. New York: Plenum; 1965. 2 volumes.

Lists approximately 8,300 organic compounds with references through 1962.

Handbook of high resolution multi-nuclear NMR. C. Brevard and P. Granger. New York: Wiley; 1981. 229 pages.

Fine tuning your NMR spectrometer. Provides 83 NMR fingerprints of nuclei, including spectra.

Handbook of NMR spectral parameters. W. Brugel. London: Heyden; 1979. 3 volumes.

"Tabulated high resolution chemical shifts and coupling constants for organic compounds according to spin system." Parameters of 7500 cmpds grouped by parent compound. "Rapid Search Index" is an alphabetical list of about 400 classes of organic compounds. Expanded and rearranged edition of the collection published in 1967, (viz.: W. Brugel, *Nuclear magnetic resonance spectra and chemical structure*).

Handbook of proton-NMR spectra and data. Ashai Research Center, Tokyo, supervised by Shinichi Sasaki. Tokyo and Orlando, FL: Academic, 1985-. 11 volumes.

Projected as an open-ended series of spectra and data. As of 1986 consists of 8000 compounds divided into 10 volumes with 800 entries each. Each volume is indexed independently by name, molecular formula, substructure and chemical shift. There is also a cumulative index for volumes 1-10.

Handbook of tritium NMR spectroscopy and applications. E.A. Evans et al. Chichester, NY: Wiley; 1985. 249 pages.

Reviews all aspects of the development and applications of ^{3}H NMR spectroscopy over 15 years of its routine use. Designed as a comprehensive handbook of information, references and basic test for users of radioisotopes in life sciences research. Includes bibliography.

NMR band handbook. Herman A. Szymanski and Robert E. Yelin. New York: IFI/Plenum; 1968. 432 pages.

Lists 1200 organic compounds; 4800 shifts. Does not include spectra. Formula and shift indexes.

NMR data tables for organic compounds. Volume 1. Frank Bovey. New York: Wiley; 1967.

4,230 compounds. Does not include spectra. Formula index.

Nuclear magnetic resonance spectra. Philadelphia, PA: Sadtler Research Laboratories; 1966-. Looseleaf.

52,000 spectra by 1990. Indexed through the *Sadtler cumulated comprehensive indexes.*

Nuclear magnetic resonance spectra and chemical structure. Werner Brugel, compiler. New York: Academic; 1967-. 235 pages.

Table of NMR parameters of compounds the spectra of which have been analyzed. Organic compounds only. Does not include spectra. Alphabetical index. Volume 1 only; updating was intended but abundance of material resulted in completely new edition in 3 volumes, viz. *Handbook of NMR spectral parameters.*

Proton and carbon NMR spectra of polymers. Quang Tho Pham, Roger Petriaud and Hugues Waton. Chichester, NY: Wiley: 1983-84. 3 volumes.

Indexed by name within broad chemical classes.

The Sadtler guide to NMR spectra. William Walter Simons and M. Zanger. Philadelphia, PA: Sadtler Research Laboratories; 1972. 542 pages.

A systematic presentation of interpreted NMR spectra, arranged by proton type . . . selected from the *Sadtler catalog of NMR spectra* either because they represent examples of characteristic absorbance patterns or because they are commonly encountered proton types. Indexed by spin system, coupling contents and exchangeable proton types as well as alphabetically.

The Sadtler guide to the NMR spectra of polymers. William Walter Simons and M. Zanger. Philadelphia, PA: Sadtler Research Laboratories; 1973. 298 pages.

Illustrates the spectra of a variety of common polymers and copolymers. Includes "Polymer Find Chart" and manufacturer, commercial name and alphabetical indexes.

Sadtler handbook of proton NMR spectra. William W. Simons, editor. Philadelphia, PA: Sadtler Research Laboratories; 1978. 1254 pages.

Arranged by functional group. Complementary to the handbooks of IR and UV spectra, containing the same compounds and using the same spectral numbering system. Alphabetical index; separate index volume also includes book order and chemical shift indexes.

Sadtler NMR spectra handbook of esters. Philadelphia, PA: Sadtler Research Laboratories; 1982. 686 and 159 pages.

The term 'esters' as used to describe the contents of this volume is applied only to the non-cyclic esters of carboxylic acids. Other types of esters will appear in future volumes dedicated to that group of compounds, e.g., sulfonic acid esters will be included with other compounds containing sulfur. 2,000 spectra selected from among those published in the Standard proton NMR collections between 1966 and 1980. Alphabetical, molecular weight and numerical indexes.

Selected nuclear magnetic resonance spectral data: 40 MHz, 60 MHz. American Petroleum Institute. Research Project 44. College Station, TX: Thermodynamics Research Center, Texas A & M; 1959-. Looseleaf.

Principally hydrocarbons and compounds of interest to the petroleum industry. 974 spectra by 1976. Index included in the 100 MHz collection; also indexed by compound name and formula in the *Comprehensive index of API 44-TRC selected data.*

Selected nuclear magnetic resonance spectral data: 100 MHz. American Petroleum Institute. Research Project 44. College Station, TX: Thermodynamics Research Center, Texas A & M; 1974-. Looseleaf.

Principally hydrocarbons and compounds of interest to the petroleum industry. 414 spectra by 1976. Indexed; index includes 40, 60 MHz collection; also indexed by compound name and formula in the *Comprehensive index of API 44-TRC selected data.*

100 MHz nuclear magnetic resonance spectra. Philadelphia, PA: Sadtler Research Laboratories; 1969-1970. 2 volumes. Looseleaf.

768 spectra. Indexed by name, chemical class and molecular formula.

Ultraviolet Spectra

Absorption spectra in the ultraviolet and visible region. L. Lang, editor. New York: Academic; 1961.

Lists approximately 6,000 organic and inorganic compounds. Includes spectra. Formula and substance indexes.

Atlas of protein spectra in the ultraviolet and visible regions. Donald Kirschenbaum, editor. New York: IFI/Plenum; 1972. 3 volumes.

Lists 1700 compounds. Includes spectra. Chemical substance index. Vol. 3: Bibliography of about 2500 citations for the 2434 indexed spectra.

Atomic and ionic spectrum lines below 2000 Angstroms: hydrogen through krypton. Raymond L. Kelly. American Chemical Society and the American Institute of Physics for the National Bureau of Standards; 1987. 3 volumes.

"Spectrum lines from the elements hydrogen through krypton, in all stages ionization, have been collected from the open literature and critically compared with the best values of atomic energy levels. The resulting list is ordered by element, spectrum number, and wavelength. The classification of each transition is given, along with the upper and lower energy level. A finding list [arranged by wavelength] is included." Issued as a supplement to the *Journal of Physical and Chemical Reference Data.*

Atomic spectra in the vacuum ultraviolet from 2250 to 1100 Å. Joseph Junkes, E.W. Salpeter and G. Milazzo. Citta del Vaticano: Specola Vaticano; 1965-.

Hollow cathode spectra. Chart of H_2 Lyman bands included.

Handbook of HeI photoelectron spectra of fundamental organic molecules: ionization energies, ab initio assignments, and valence electronic structure for 200 molecules. K. Kimura et al. Tokyo: Japan Scientific Societies Press and New York: Halsted; 1981.

Emphasis on substitution effects. Includes some simple inorganic molecules important as starting molecules for interpretation.

Handbook of ultraviolet and visible absorption spectra of organic compounds. Kenzo Hirayama. New York: Plenum; 1967. 642 pages.

Lists 8,443 compounds. Does not include spectra. No formula or substance indexes.

Organic electronic spectral data. New York: Interscience; 1946-.

Cooperative effort to abstract and publish in formula order all the UV-visible spectra of organic compounds presented in the journal literature from 1946 onward. Entries include solvent or phase and wavelength values. Organized according to molecular formula index system used by Chemical Abstracts. Does not include spectra. Large publication gap: volume 25, covering literature through 1983 was copyrighted 1989.

Sadtler handbook of ultraviolet spectra. Philadelphia, PA: Sadtler Research Laboratories; 1979.

Abridged edition of the Sadtler standard UV spectra collection intended as a small convenient collection of reference spectra of organic compounds relevant to introductory courses in organic chemistry, or as a desk reference when comprehensive collections are not available. Over 2,000 spectra representing the neutral, base and acid scans for 1600 compounds. Complementary to the handbooks of IR and NMR spectra, containing the same compounds and using the same spectral numbering system. Alphabetical, numerical and locater indexes.

Selected ultraviolet spectra data. American Petroleum Institute. Research Project 44. Washington, DC: American Petroleum Institute; 1945-.

Principally hydrocarbons and compounds of interest to the petroleum industry. 1178 spectra by 1976. Indexed; also indexed by compound name and formula in the *Comprehensive index of API 44-TRC selected data.*

Ultraviolet spectra. Philadelphia, PA: Sadtler Research Laboratories; 1960-. Looseleaf.

41,829 spectra through 1989. Indexed through the *Sadtler cumulated comprehensive indexes.*

Ultraviolet and visible absorption spectra: index, 1930-1963. H.M. Hershenson. New York, London: Academic; 1966. 3 volumes.

Lists bibliographic references for spectra of 73,000 organic and inorganic compounds. Chemical substance index.

UV atlas of organic compounds (the DMS-UV atlas). London: Butterworths; 1966-71. 5 volumes.

Lists approximately 1,000 compounds. Includes spectra. Indexes.

Miscellaneous and Various Spectra

Assignments for vibrational spectra of seven hundred benzene derivatives. Gyorgy Varsányi. New York: Wiley; 1974. 2 volumes.

700 compounds (IR and Raman spectra). Spectra in volume 2.

Atlas of electron spin resonance spectra. Ya. S. Lebedev et al. Akademiia nauk SSSR. Institut khimicheskoi fiziki. New York: Consultants Bureau; 1963-1964. 2 volumes.

Subtitle: Theoretically calculated multicomponent symmetrical spectra. Author translation from Russian original.

Atlas of near infrared spectra. Philadelphia, PA: Sadtler Research Laboratories; 1981. ~999 pages.

Collection of reference NIR spectra of 1,000 compounds selected on the basis of their demand by research scientists, their value for

NIR methods development and their general usefulness in industry. Alphabetical and molecular formula indexes.

Atlas of spectral data and physical constants for organic compounds. 2nd edition. Jeanette Grasselli and William Ritchey, editors. Cleveland: CRC Press; 1975. 6 volumes.

Lists 21,000 compounds. Does not include spectra. Indexes and bibliographies.

Comprehensive index of API 44-TRC selected data on thermodynamics and spectroscopy. College Station, TX: Texas A & M University. Thermodynamics Research Center; 1968, 1974.

Includes compound name and formula indexes to the American Petroleum Institute Research Project 44 collections or selected spectral data.

CRC handbook of EPR spectra from quinones and quinols. Jens A. Pedersen, editor. Boca Raton, FL: CRC Press: 1985. 382 pages.

Compilation of EPR data of numerous semiquinones and related radicals. Some spectra included as examples; most data presented in tabular form. Compound index and subject index.

Données spectroscopiques relatives aux molecules diatomiques: spectroscopic data relative to diatomic molecules. B. Rosen. Oxford, New York: Pergamon; 1970. 515 pages.

"Completely recast" updating of the 1951 edition. The tables mainly deal with electronic spectra located in the infrared, the visible and the ultraviolet.

Electronic absorption spectra of radical ions. Tadamasa Shida. Amsterdam and New York: Elsevier; 1988. 446 pages.

Charts for approximately 700 ions of spectra produced in gamma-irradiated frozen solutions. Indexed by name and by molecular formula.

Handbook of Auger electron spectroscopy. 2d edition. Lawrence E. Davis et al. Eden Prairie, MN: Physical Electronics Industries; 1976. 252 pages.

Includes 115 spectra arranged by atomic number, introductory chapters on method, and alphabetical index.

High resolution spectral atlas of nitrogen dioxide 559-597 nm. Kiyoji Uehara and H. Sasda. Berlin and New York: Springer-Verlag; 1985. 225 pages.

"The major part of the book consists of an atlas..of the absorption spectrum and the Stark modulation spectrum of NO_2 measured using a cw dye laser." — Preface. Chapter 4 is a list of the literature of spectroscopic studies of the NO_2 molecule published between 1978 and the beginning of 1984.

Inductively coupled plasma-atomic emission spectroscopy; an atlas of spectral information. R.K. Winge et al., editors. Amsterdam and New York: Elsevier; 1985. 584 pages.

Presents 232 wavelength scans of 70 elements and 973 prominent lines with associated detection limits for 72 elements. Each section is arranged by chemical symbol.

Line coincidence tables for inductively coupled plasma atomic emission spectrometry. P.W.J.M. Boumans. 2d edition. New York and Oxford: Pergamon; 1980. 2 volumes.

The tables concern spectral interferences emitted by free atoms or ions of the concomitants of the sample. The principal aim is to aid line selection in any type analysis. 896 analysis lines of 67 elements are listed. Arranged alphabetically by symbol.

Raman/IR atlas organischer Verbindugen//Raman/IR atlas of organic compounds. (DMS) Weiheim: Verlag Chemie; 1978. 2 volumes.

Raman and IR spectra, reproduced together on the same page in each case, for about 1,000 compounds. Divided into 15 main groups and 93 classes. Indexed alphabetically, by formula, and by substituent.

Selected Raman spectral data. American Petroleum Institute. Research Project 44. Washington, DC: American Petroleum Institute; 1948-. Unpaged.

Principally hydrocarbons and compounds of interest to the petroleum industry. 652 spectra by 1976. Indexed; also indexed by compound name and formula in the *Comprehensive index of API 44-TRC selected data.*

Spectral atlas of nitrogen dioxide, 5530 to 6480Å. Donald K. Hsu, David L. Monts and Richard N. Zare. New York: Academic; 1978. 632 pages.

Chapter 1 is a line atlas of high resolution absorption spectra presenting approximately 19,000 prominent lines; Chapter 2 is a collection of rotational analyses in the region from 5700 to 6800Å; and Chapter 3 is an annotated bibliography.

Spectral atlas of polycyclic aromatic compounds. W. Karcher et al., editors. Dordrech and Boston: D. Reidel; 1985. 818 pages.

Provides several types of spectra for each of 42 PAC pollutants. Additional data supplied for each compound include sources, distribution, and biological effect (e.g., carcinogenic activity). Arranged by mass number; indexed by compound name.

Spectral data of natural products. Kazutaka Yamaguchi. Amsterdam, New York: Elsevier; 1970.

Lists data for IR, UV, NMR, mass spectra, optical rotatory dispersion and circular dichroism values for 4,350 compounds. Occasionally includes spectra. Chemical substance index.

Spectroscopic data. 1975-76. J.E. Melzer and S.N. Suchard. New York: IFI/Plenum. 2 volumes in 3.

Collection of spectroscopic information dealing primarily with the electronic spectra of diatomic systems. Does not include spectra. Volume 1 parts 1-2: Heteronuclear diatomic molecules (restricted to molecules that, from thermodynamic reasoning, could be produced in an electronically excited state by a chemical reaction between a ground state atom and a ground state molecule). Volume 2: Homonuclear diatomic molecules.

Spectroscopic references to polyatomic molecules. V.N. Verma. New York: IFI/Plenum; 1980. 126 pages.

Bibliography of about 900 organic ring compounds; arranged in alphabetical order.

Tables for emission spectrographic analysis of rare earth elements. Ch. Kerekes. New York: Macmillan; 1964. Unpaged.

Analysis lines for each rare earth element (and scandium and yttrium) given in order of decreasing wave length. Arranged by chemical symbol.

Tables of spectral data for structure determination of organic compounds. 2d edition. New York and Berlin: Springer-Verlag; 1989. 1 volume, various paging.

For organic chemists involved in the identification of organic compounds or the elucidation of their structure. Arranged by type of spectra. Translated from the 3rd German edition. Includes tables for ^{13}C-NMR, ^{1}H-NMR, IR, mass, and UV/VIS, as well as a section of summaries arranged by structural elements and a section containing rules of practical importance.

Tables of spectral lines. Aleksandr Zaidel et al. New York: IFI/Plenum; 1970. 782 pages.

Part 1 contains spectral lines of 60 elements in order of decreasing wavelength. Part 2 gives spectral lines of 98 elements individually for each. Part 3 contains auxiliary reference material.

NOTES

1. Paramagnetic species are those containing atoms with unpaired electrons.
2. *Beilsteins Handbuch der organischen Chemie.* Heidelberg, Germany: Springer-Verlag; 1881-.
3. *Gmelins Handbuch der anorganischen Chemie.* Frankfort, Germany: Gmelin Institute; 1817-.

NEW REFERENCE WORKS
IN SCIENCE AND TECHNOLOGY

Arleen N. Somerville, Editor

Reviewers for this issue are: Julie M. Hurd (JMD), University of Illinois at Chicago; Isabel Kaplan (IK), University of Rochester; Donna Lee (DL), University of Vermont; Diane J. Reiman (DJR), University of Rochester; Bruce Slutsky (BS), St. John's University; Arleen N. Somerville (ANS), University of Rochester; Jack W. Weigel (JWW), University of Michigan.

GENERAL SCIENCE

The almanac of science and technology, what's new and what's known. Edited by Richard Golob and Eric Brus. Boston: Harcourt Brace Jovanovich; 1990. 531 p. ISBN 0-15-600049-4.

This is not a review of the year's developments in science, nor is it an in-depth history of science. Rather, it is truly an almanac, an easily readable account of current topics in nine general areas: astronomy, biology, brain and behavior, chemistry, computers, earth sciences, environment, medicine, and physics. There is an index, but no references and no bibliographies for further reading. The copy under review is marked "uncorrected proof," which may account for the generally poor quality of illustrations and occasional editors' correction marks on some figure numbers. Despite these cosmetic flaws, this is an interesting book to read for general awareness of what's happening in science. Chapters are each approximately sixty pages and cover about six topics. While not recommended for research collections, this book could prove quite a handy addition to a public or community college library, or to the leisure reading room of a university library. (IK)

Research centers directory. 15th ed. New York: Gale; 1990. 2 vol. $415. ISBN 0-8103-7350-5.

The directory describes 12,326 non-profit research centers in the hard and soft sciences, located in North America. Entries detail the parent institution and other supporting organizations, staff, research activities, publications, educational activities, public services, and the extent of their libraries, along with phone, fax, address, and director. The indexes and the organization of the publication provide many points of access including, broad subject (17 of them), controlled subject (about 5,000), title of the center, projects, programs, parent institutions, and keyword. At the time this review was written, Gale was in the process of preparing the directory to be mounted as a database on Dialog. Libraries with little need for this info could rely on Dialog access. Libraries affiliated with research or marketing organizations will want to determine how quickly the fees for Dialog searches will add up to the price of the publication. (DL)

Salaries of scientists, engineers and technicians . . . a summary of salary surveys, 14th ed. Prepared by Eleanor L. Babco. Washington, DC: Commission on Professionals in Science and Technology; 1990. 230 p. $50.00.

This work is a compendium of tables of salary data on personnel in science and technology, grouped as follows: starting salaries, salaries of experienced scientific and technical personnel, salaries of engineers, salaries of technicians, and faculty salaries. Within categories there are numerous tables specific to discipline (chemistry, geology, engineering, math, etc.) and with differing parameters, such as degree level, geographic area, type of employer, sex, job function, etc. The source of the data appears on each table. Full bibliographic information is given at the end of the volume in "Bibliography of sources," a useful listing of government, commercial and academic organizations which survey and compile data on salary and employment. Recommended for libraries, career counseling services and university administrators. (IK)

Science and engineering indicators—1989. Prepared by the National Science Board. Washington, DC: U.S. Government Printing Office; 1989. 415 p.

This is the ninth in the series of biennial *Science Indicators* reports produced by the National Science Board, as mandated by the National Science Foundation Act of 1950. The report presents, through text, tables and graphs, information on the current status of and future trends in U.S. and foreign science, including support for research endeavors; funding and coursework for precollege and higher education; characteristics of the labor force; indicators of innovation, such as patents, small business start-ups and success; industrial productivity and trade; and public attitudes toward science and technology. The Appendix Tables, which comprise about half the book, show data from which many of the graphs and textual comments were derived. Data from other sources are noted at the bottom of figures or as footnotes to the text. Educators, public policy analysts, and corporate planners will find this a valuable resource. Recommended. (IK)

EARTH SCIENCES

Encyclopedia of igneous and metamorphic petrology. Edited by D.R. Bowes. (Encyclopedia of Earth Sciences Series) New York: Van Nostrand Reinhold; 1989. 666 p. $112.95. ISBN 0-442-20623-2.

This is another in the valuable Encyclopedia of Earth Sciences Series, which seeks to provide reference works useful to earth scientists, and also to scientists in other disciplines who need ready access to the information. This objective is met exceedingly well with this volume on petrology. Entries are written clearly and define terms that may not be known to readers outside the specialty. Each entry includes a bibliography of important books and journal articles and an often extensive list of cross references to other entries. Illustrations, graphs, and charts are used liberally. Recommended for all collections that serve clientele who need information about petrology. (ANS)

Dictionary of paleontologists of the world. 5th ed. By Rex A. Doescher. Washington, DC: International Paleontological Association; 1989. 447 p. $20.00. No ISBN given.

This important directory provides the following information about paleontologists from all countries: name, address, phone number, and subject specialties. The Alphabetic Taxonomic Index lists paleontologists under their specialties. The useful Geographical Index lists each institution alphabetically by country, state or province, city, address, and the paleontologists within each institution. As was true for the 4th edition, China and the Soviet Union paleontologists are included (seven and one half pages of Chinese scientists and seven pages of Soviet scientists). Essential for all paleontological research collections. (ANS)

The concise Oxford dictionary of earth sciences. Edited by Ailsa Allaby and Michael Allaby. Oxford, New York: Oxford University Press; 1990. 410 p. $39.95. ISBN 0-19-866146-0.

This *Dictionary* joins two other recent geology dictionaries: The 1987 *Glossary of Geology* from the American Geological Institute (3rd edition) and the 1986 *Challinor's Dictionary of Geology* (6th edition, Oxford University Press). *Challinor's Dictionary* provides very extensive definitions for fewer entries in its 344 pages. The AGI *Glossary* includes more entries, often more extensive than the Oxford *Dictionary*, and offers a more extensive bibliography in the "list of references cited" in its 750 pages. The Oxford *Dictionary* provides brief, concise definitions of terms from earth sciences defined broadly. Brief biographic notes about world-wide earth scientists of historic importance are unique among the three dictionaries. Occasionally the Oxford *Dictionary* provides more extensive information than the AGI *Glossary*, such as for brachiopoda. Extensive cross references appear in the oxford *Dictionary* as well as a 10-page bibliography of major earth sciences books. Recommended for research collections that require comprehensive coverage. (ANS)

ENGINEERING AND TECHNOLOGY

Handbook of adhesives, 3d ed. Edited by Irving Skeist. New York: Van Nostrand Reinhold: 1990. $95.00. ISBN 0-442-28013-0.

The *Handbook* contains forty-seven chapters on various aspects of the physics, chemistry and applications of adhesives. Contributors are mainly industry specialists. There are five chapters on fundamentals, 28 on different types of adhesive materials (animal glue, nitrile rubbers, acrylics, amino resins, high temperature organics, etc.) and 14 on adherends and bonding technology (bonding to plastics, sealants and caulks, pressure-sensitive adhesives, conductive adhesives, adhesives in the aerospace industry, robotic dispensing of adhesives, etc.). There are many tables, diagrams, and charts. Chapters conclude with a bibliography; some are further augmented by lists of additional readings, standards, and commercial producers of materials. Recommended. (IK)

Handbook of compressed gases. 3rd. ed. (Compressed Gas Association) New York: Van Nostrand Reinhold; 1990. 657 p. $72.95. ISBN 0-442-21881-8.

This *Handbook* focusses on the properties and accepted methods of storage, handling, and transportation of compressed gases. It is intended to serve a wide audience: workers, managers, scientists and engineers, teachers, and laypeople. The first part provides general information about gases: properties, safety guidelines, and regulations. The second part describes technical information about 45 specific gases and gas mixtures. This information includes: name synonyms, Chemical Abstracts Service Registry Number, molecular formula, relevant physical constants, chemistry, uses, physiological effects, safe storage and handling and use, disposal, handling of leaks and emergencies, shipping methods, recommended containers, manufacturing methods, and costs. Recommended for all public, academic, and industrial libraries whose clientele use compressed gases. (ANS)

Directory of Japanese technical resources in the United States, 1990. (Japanese Directories series.) Prepared by the U.S. Department of Commerce. For sale from the National Technical Information Service. 158 p. $39.00. ISBN 0-934213-27-5.

The bulk of this work is the section,"Directory of Information Services," which covers "over 250 commercial services, government agencies and libraries which collect, abstract, translate or disseminate Japanese technical information." Data include address, phone and fax numbers, services offered, publications/databases of the organization, and contact person. The directory is indexed by expertise, geographic location, and services provided. Following these is a "Directory of Federally Funded Translations," a listing of about 135 Japanese technical documents translated at US government expense from January 1989 to March 1990. Information is in GRAI format, with GRAI number, NTIS order numbers and price codes. There are two companion directories to the volume under review, both produced by NTIS, *Directory of Japanese Databases – 1990* ($35) and *Directory of Japanese Technical Reports 1989-90* ($35). The price for the set of three volumes is $89. (IK)

HEALTH SCIENCES

Antiepileptic drugs. 3rd Edition. Edited by Rene H. Levy. . . . [et al.]. New York: Raven Press; 1989. 1025 p. $135.00. ISBN 0-88167-539-3.

This third edition demonstrates the better use and testing of antiepileptic drugs through the applications of advances in basic neuroscience and new knowledge from neuropharmacology and pharmacokinetics. Using the same format of the previous editions it points the way to improved care of patients with seizure disorders. The first section considers general topics such as biotransformation, toxicology, mechanisms of action, drug selection, and discontinuation of therapy. Later in the text the contributing authors review chemical and pharmacological data for phenytoin, phenobarbital, primidone, carbamazepine, valproate, ethosuximide, the benzodiazepines, and other drugs. Extensive bibliographies with full citations and a detailed subject index are provided. Essential for special libraries of pharmaceutical companies and universities with programs in the pharmaceutical sciences. (BS)

Dictionary and handbook of nuclear medicine and clinical imaging. By Mario P. Iturralde. Boston: CRC; 1990. 564 p. $125. ISBN 0-8493-3233-8.

Procedures for and results of nuclear medicine and radiology impact on health professionals in many fields. The over 500 definitions found in this dictionary are surprisingly simple for such a complex field. Some terms are unique, but many appear in everyday speech, with quite different meanings. A bead, for instance, is "a small part of a program written to perform a specific job. Beads can be developed individually and then strung in 'threads' to form a complete program." Even specialists may welcome the establishment of precise meanings for those terms which have yet to appear in any dictionary. Students of nuclear medicine and radiology will also want access. A good example of what a specialty dictionary should be. The second half of the work is devoted to tables outlining the chemical properties of substances used in imaging. Other tables list measurements and calculations. (DL)

Drug dosage in laboratory animals: a handbook. 3rd rev. and enlarged ed. By R.E. Borchard, C.D. Barnes, and L.G. Eltherington. Caldwell, NJ: Telford Press; 1990. 692 p. $69.50. ISBN 0-936923-19-9.

Most of this source is devoted to tables of over 7300 doses for over 440 drugs commonly used in biomedical research. The drug responses, obtained from the published literature are classified according to toxicity, primary use or action, and secondary use or action. The compilers chose drugs and investigational agents used in several animal species and via more than one route of administration. Researchers may add information from their own experiments to the tables. The introductory material discusses factors which modify drug responses and anesthesia. This third edition includes 230 new drugs, 1000 additional references, and 2700 new doses. For libraries supporting research in experimental medicine. (BS)

Drug-test interactions handbook. Edited by J.G. Solway. London: Chapman and Hall; 1990 (Published in the USA by Raven Press) 1087 p. $325.00. ISBN 0-88167-602-0.

Problems that occur when drugs and other factors affect particular clinical laboratory tests necessitated the publication of this source. It is divided into three parts covering blood specimens, urine specimens, and clearance studies, respectively. Within each part the tests are divided into groups such as bone and joint diseases, kidney function tests, and thyroid function tests. The matrices summarize the results for each drug-test or factor-test interaction and refer users to the entry that summarize the results of the test. A literature reference is given for each entry. Highly recommended for collections in clinical chemistry. (BS)

Drugs available abroad. Edited by Jerry L. Schlesser. New York: Gale; 1991. 410 p. $89.95. ISBN 0-8103-7177-4.

What took health sciences publishers so long to come up with this reference source? It covers more countries and gives more data on each drug than *Index Nominum*. In addition to getting you from the proprietary to the generic name, entries list other synonyms, manufacturers, drug actions, indications, the form in which the drug is dispensed, dosage, contraindications, interactions, adverse effects, and the drug supplied in the US for the same indications, when applicable. The international listings were obtained by Derwent, Inc., which consulted their own databases of drug information in addition to the drug compenduims and pharmacopeias of the countries covered. Western European countries and

Mexico are the most widely represented. Gale plans to publish new editions annually. Indexes are good. The addition of pictures as found in the *PDR* would be helpful, but we can't have everything in a first edition. Given the surprisingly reasonable price, every medical, hospital, and pharmacy library should have access to this publication. (DL)

Goodman and Gilmans the pharmacological basis of therapeutics. 8th Edition. Edited by Alfred Goodman Gilman. . . . [et al.]. Pergamon Press. 1990. 1811 p. $85.26. ISBN 0-08-040296-8.

Like its predecessors, this edition correlates pharmacology with related medical sciences, interprets the actions and uses of drugs from the viewpoint of recent medical advances, and emphasizes the applications of pharmacodynamics to therapeutics. It is reorganized to include a chapter on the Principles of Toxicology in the section on General Principles, newly authored chapters in the section on Autonomic Pharmacology, and new sections on drugs that affect gastrointestinal functions, immunosuppressive agents, and dermatopharmacology. Over 50 new drugs and investigational agents such as AZT, tissue plasminogen activator, erythropoietin, and lovastatin are discussed. The editors provide literature references with full citations, an appendix with pharmacokinetic data on 243 drugs, and a detailed subject index. Essential for libraries with collections in the pharmaceutical sciences. (BS)

Handbook of family measurement techniques. By John Touliatos et al. London: Sage; 1989. 797 p. $65. ISBN 0-8039-3121-2.

976 questionnaires and other measurement instruments are succinctly described in entries of one page or less. Sample items from the instrument are provided, if the publisher has consented. The final paragraph for each entry summarizes the reliability and validity of the test. Entries conclude with brief lists of references. The description of the tests are not as detailed as those found in the *Mental Measurements Yearbook,* but, as you might expect, this source covers a larger number of instruments related to the study of the family. (DL)

Handbook of pharmacokinetics: toxicity assessments of chemicals. English ed. By Jean-Pierre Labaune [translated by G. Tyas and C. Ioannides]. Chichester: Ellis Horwood; New York: Halsted Press; 1989. 704 p. $65.00. ISBN 0-7458-0256-7 (Ellis Horwood Limited) ISBN 0-470-21572-0 (Halsted Press).

Labaune introduces readers to all aspects of pharmacokinetics including absorption, distribution, bioavailability, metabolism, and the development of dose regimens. His work is suitable as a reference for pharmacologists, toxicologists, biochemists and medical scientists and as a textbook for a course at the advanced undergraduate or first-year graduate level. It emphasizes practical aspects and is limited to the study of drug disposition after oral or intravenous administration. Readers will gain enough expertise to design appropriate experimental studies as part of the safety evaluation of chemicals and drugs. *Disposition of toxic drugs and chemicals in man* (Year Book Medical Publishers) is a good alternative source when pharmacokinetic data on a specific drug is needed. (BS)

Long term care standards manual, 1990. Chicago: Joint Commission on Accreditation of Healthcare Organizations: 1989. 212 p. $50. ISBN 0-86688-208-1.

Since the last edition of this manual, the Joint Commission has revised their standards in the areas of Quality Assurance, patient complaints, and pharmaceutical activities. Even unaccredited long term care facilities will want this publication, if only as a guide for self evaluation. Real estate developers and community services for the elderly may also be interested in recommendations regarding the physical plant. (DL)

The Merck manual of geriatrics. Edited by William B. Abrams and Robert Berkow. Rahway, NJ: Merck; 1990. 1267 p. $17.60. ISBN 9110910-32-8.

The Merck Manual, the world's most widely used medical text (according to their Forward, anyway), is a reliable source for standard, accepted explanations of etiology, procedures for diagnosis, and therapy recommendations for all medical disorders. In order to fulfill this mission it also covers normal human development and activity. But etiology, diagnostic procedures, therapeutic actions, and the very definition of normal can change as people age. This new publication offers suggestions, specific to elder care, for history taking, diagnosis, therapeutic objectives, preventive medicine, sources of care, health technology, surgery, rehabilitation, exercise, terminal care, and legal and ethical issues. The chapters have a similar arrangement to those in the original *Merck Manual,* but have been completely rewritten to concentrate on issues relevant to geriatrics. The new manual also includes considerably less technical detail. These manuals resemble reference books rather than textbooks. This arrangement makes the individual facts in these manuals more accessible to readers who just want an answer, not a lecture. Recommended for academic health care and hospital collections, and public libraries with a substantial elderly clientele. (DL)

LIFE SCIENCES

Animal anatomy on file. By the Diagram Group. New York: Facts on File; 1990. 200 plates. $145. ISBN 0-8160-2244-5.

The editors who compiled this notebook envision three uses for it: as a guide, a copy file, and a source for exams or self tests. But these diagrams do not convey enough information on their own to serve as a complete guide to animal anatomy. On the other hand, the simplicity of their design makes them an ideal source of illustrations for lectures, syllabi, and other instructional materials, including examinations. The publisher states explicitly that these drawings may be used for non-profit, educational purposes. The index is arranged by organism. (DL)

Handbook of enzyme inhibitors. 1st ed. By Helmward Zollner. New York: VCH; 1989. [Reprinted 1990] 440 p. $126.00. ISBN 0-89573-860-0.

This source with listings for over 5,000 inhibitors for about 1,000 enzymes allows users to search for an inhibitor of a particular enzyme and for the enzymes which are inhibited by a specific inhibitor. This information contributes to the understanding of metabolic processes and of enzyme mechanisms. Biochemists, biologists, pharmacists, toxicologists, and enzymologists can use this handbook to help plan investigations and interpret the experimental results. The entries include the enzyme name (recommended by the International Union of Biochemistry), inhibitor name (recommended by the Merck Index), type of inhibition, effect concentration, comments, and literature references. For biochemistry collections. (BS)

The Human body on file. By the Diagram Group. New York: Facts on File: 1983. 300 plates. $145. ISBN 0-87196-706-5.

Like *Animal Anatomy on File,* this publications offers illustrations designed for photocopying or tracing. Because of the simplicity of the diagrams it is not recommended as an atlas of human anatomy. The index of body parts is very complete. (DL)

Sigma-Aldrich handbook of stains, dyes, and indicators. By Floyd J. Green. Milwaukee, WI: Aldrich Chemical Co.; 1990. 776 p. $70.00. ISBN 0-941633-22-5.

This handbook is an up-to-date, concise, and easily accessible guide to 525 compounds used as stains, dyes, and indicators. Most of the compounds are listed in the Aldrich Catalog or Sigma Chemical Co. catalog. The compounds are listed alphabetically by common name. Each entry includes a description of dye type and use; technical data such as grade; Chemical Abstracts Service Registry No.; Colour Index no.; molecular formula; molecular weight; melting and boiling points; appearance; solubilities in water, ethanol, and methyl Cellosolve; history; preparation; synonyms; references to other major sources and spectra when easily available. Name, Registry no., and Colour Index no. indexes facilitate rapid access. Recommended for all collections serving scientists with a need for data about stains and dyes. (ANS)

MATHEMATICAL SCIENCES

Archives of data-processing history: a guide to major U.S. collections. Edited by James W. Cortada. New York, NY: Greenwood Press; 1990. 181 p. $45.00.

In the history of science and technology, the history of computing has only recently attracted scholars and publishers. This book is meant to serve as a quick guide to significant collections of primary sources in the United States. The editor acknowledges in his preface that many more collections are likely to become accessible in the next decade as institutions recognize the potential value of the materials. Twelve contributed articles describe the archival holdings of governmental units, universities, corporations, museums, and centers such as the Charles Babbage Institute for the History of Information Processing, a leader in archival activities. The editor completes the volume with a concise bibliographical essay on U.S. archival holdings. Clearly, sufficient raw materials await the examination of historians of data-processing, and this book enhances access to important resources. Highly recommended for research collections in computer science, and the history of science and technology. (DJR)

PHYSICAL SCIENCES

Comprehensive medicinal chemistry; the rational design, mechanistic study, and therapeutic application of chemical compounds. Chairman of the Editorial Board: Corwin Hansch; Joint Executive Editors: Peter G. Sammes and John B. Taylor. 6 vols. Oxford, New York: Pergamon; 1990. $1995.00. ISBN 0-08-0325-30-0 (set).

The structure, structure-activity relationships, synthesis and design, chemistry, reactions, and pharmacological activities of medicinal substances are discussed comprehensively. Extensive references throughout the text refer to original articles. In most volumes the information is organized by function of the compound. Volume 2, for example, covers enzymes that serve as agents acting in various functions, such as those acting on nucleic acids. This volume includes an extensive discussion of beta-lactam antibiotics. Volume 3 covers receptors, such as neurotransmitter and autocoid receptors. This section includes a fifty page discussion of dopamine, with 144 references. Volume 4 is devoted to drug design. A good summary of orphan drugs is one of the specific topics grouped together in volume 1. Each chapter is written by recognized experts. The subject index in volume 6 is detailed and extensive. The "Drug Compendium" in volume 6 provides very brief information for over 5500 medicinal substances: chemical structure, name, molecular formula, type of drug, and Chemical Abstracts Service Registry Number. There are references to the other 5 volumes, if the substances are covered, and a couple key references. A Classification Index lists substances by pharmacological activity. Recommended for libraries serving medical, pharmacology, and synthetic organic chemistry researchers if funds permit. (ANS)

CRC handbook of ion-selective electrodes: selectivity coefficients. Yoshio Umezawa, Editor. Boca Raton: CRC Press; 1990. 848 p. $295.00. ISBN 0-8493 0546-2.

This handbook lists selectivity coefficients for over 1600 electrodes along with their electrode characteristics. The information is arranged alphabetically under four headings: inorganic anions, organic anions, organic cations, and inorganic cations. The data is extracted from articles published in major journals from 1966 through summer of 1988. Recommended for comprehensive analytical chemistry collections. (ANS)

Dictionary of Astronomical Names. By Adrian Room. London and New York, Routledge, 1988, 282 p. $27.50. ISBN 0-415-01298-8.

This compilation begins with a 27-page review of the history and principles of astronomical nomenclature. There follows a brief "astronomical glossary" which defines basic, general terms such as *asteroid* and *refractor.* The "dictionary" portion lists the proper names of several hundred selected individual astronomical objects — stars, minor planets, constellations, planets, planetary satellites, and lunar features — along with a few descriptive or historical lines about each. Appendix I provides a brief essay on lunar nomenclature and a selected list of the names of lunar craters with the full names, dates, nationalities, and professions of the individuals whose surnames were applied to the craters. Appendix II gives a bit of background information on minor planet nomenclature and a list of the first 1000 minor planets to be discovered (out of approximately 3000 that have been discovered and numbered). There are indications much of the data were derived from primary sources, some of which are more comprehensive than the present work. The author, evidently not a professional astronomer, is a prolific compiler of dictionaries on sundry other subjects. This book is a useful quick-reference tool which is likely to be useful particularly in libraries lacking a complete collection of more comprehensive publications. (JWW)

Dictionary of Drugs. J. Elks and C.R. Ganellin, Editors. 2 vols. London: Chapman and Hall; 1990. $1099.00. ISBN 0-412-27300-4 (set).

This two volume set serves as a definitive source of concise data about significant drugs. It succeeds very well. The format is similar to the well known *Dictionary of Organic Compounds (DOC).* Entries are arranged alphabetically by name. Entries include: common names, Chemical Abstracts names linked to relevant Collective Indexes, Chemical Abstracts Service Registry Number, structure diagram, molecular formula, physical and chemical properties for the parent compound and biologically active derivatives, therapeutic uses, hazard information, and references to major journal articles, patents, and reference books for preparation, spectra, toxicity, and biological activity.

The *Dictionary* includes substances that have significant pharmacological interest, including new drugs. Some entries are based on information in the *DOC*, but are substantially updated. A comparison of the adriamycin entries verifies this. The brief three inch entry for adriamycin in the 1982 *DOC* is extended to twenty inches. This expanded entry includes data about fourteen biologically active derivatives and includes twenty-six literature references, compared to four in 1982. The second volume includes the following indexes: name, molecular formula, Chemical Abstracts Service Registry Number, type of compound, and structure. Headings in the Compound Type index are based on pharmacological activity (anticonvulsants, immunostimulants, antineoplastic agents) or by compound type (tetracycline antibiotics, sulfonamides). Recommended for academic, medical, public, and industrial libraries which serve researchers in synthetic organic, pharmacology, and medicine. (ANS)

Guidebook to organic synthesis. 2nd ed. By Raymond K. Mackie, David M. Smith, and R. Alan Aitken. New York: Longman Scientific & Technical; copublished with Wiley; 1990. 387 p. $34.95. ISBN 0-470-21568-2 (Wiley).

This *Guidebook* is intended for students who are beginning a serious study of organic synthesis. It is a valuable concise summary of major synthetic methodology. The information is organized by broad type of reaction, such as formation of carbon-carbon bonds, ring closure and opening, oxidation, protective groups, and specific reagents. New chapters were added for asymmetric synthesis and selenium reagents. While the entire text was updated, the chapters on silicon and phosphorus reagents were expanded substantially. Specific name reactions such as Diels-Alder and Reformatsky and reaction types such as cycloaddition can be located readily using the subject index. The six pages that summarize the Diels-Alder reaction is a concise description, largely through detailed reaction schemes, of the reaction mechanisms with examples of major synthetic applications. The bibliography of further reading leads readers to selected books and journal articles. An essential book for all college and university collections. (ANS)

Handbook of natural products data. Volume 1. Diterpenoid and steroidal alkaloids. By Atta-ur-Rahman. Amsterdam, Oxford, New York, Tokyo: Elsevier; 1990. 962 p. $353.75. ISBN 0-444-88173-5.

This volume is the first of a new series that will bring together from scattered journal sources spectral data on natural products. These complex molecules are of interest because they offer a readily available source of various structural features which may be tested for biological activity. The first series volume published covers the diterpenoid and steroidal alkaloids. Alkaloids are naturally occurring bases found in plants. The diterpenoids are those compounds with the molecular formula $C_{20}H_{32}$; steroidal alkaloids are those containing the perhydrocyclopentenophenanthrene nucleus.

This work includes data for 971 compounds as reported in the literature through 1988. The compounds listed are arranged according to their structural types in order of increasing molecular weight. For each compound the author provides a structural diagram, molecular formula, molecular weight, melting point, and Chemical Abstracts Service Registry Number. The scientific name of the plant(s) in which the compound occurs are listed with references to the original literature. UV, IR, $1H$-NMR, and $13C$-NMR spectral data are listed in descending order of magnitude. Older names for compounds and spectra for simple derivatives are also reported. End-of-the-book indexes provide access to the data by compound name, molecular formula, molecular weight, plant source and compound-type. A decision to purchase this series represents a major commitment of resources, but for those research collections with strength in the area of natural products chemistry this work should prove to be a time-saving reference and a first place to search for spectral data on natural products. (JMH)

Handbook of Space Astronomy and Astrophysics. By Martin V. Zombeck. 2nd ed. Cambridge & New York, Cambridge University Press, 1990. 440 p. $75.00 (hardcover); $34.50 (paperback). ISBN 0-521-34550-2 (hardcover); ISBN 0-521-34787-4 (paperback).

Emphasizing equations, tabular data, and graphs of observational findings, this handbook contains very little verbal explanation. Besides an opening chapter of "general data" and a concluding chapter on astronomical catalogs, there are seventeen more specialized chapters on astronomical and astrophysical subfields plus auxiliary subjects such as mathematics and statistics. At the end of each chapter there is a very brief list of source publications. There are occasional places where the data are not as current as possible. For example it is disappointing that a book with a 1990 imprint date reproduces a table of natural satellites dating from 1983; hence, numerous satellites discovered by the Voyager spacecraft are not included. On the other hand there has been a substantial increase in the amount of material; the second edition contains 114 pages more than the first edition. Again, this is an assemblage of fairly standard information in a handy single-volume package and should prove to be a rather useful quick reference. (JWW)

SCI-TECH ONLINE

Ellen Nagle, Editor

DATABASE NEWS

TOXLINE Added by DIALOG

DIALOG customers now have access to *TOXLINE*, a comprehensive collection of data from the National Library of Medicine. The database covers the adverse effects of chemicals, drugs, and physical agents on living systems. *TOXLINE* is composed of sixteen subfiles as well as an exclusive subfile, providing " a more complete form of *TOXLINE* than is available on any other online service," according to DIALOG. Of the 120,000 records added each year, approximately 75 percent are from the *TOXBIB* file subfile which is derived from *MEDLINE*.

The database covers a complete range of subjects related to toxicology including adverse drug reactions, environmental pollution, and occupational hazards. Data coverage for several of the subfiles extends back to the 1950's and 1960's; the *TOXBIB* portion begins with records added in 1965. Thus, *TOXLINE* is a key resource for current and historical toxicological data. In addition to bibliographic information, most records also contain abstracts. The database may be searched using words from the abstract, descriptor, check tag, and title fields. Searches can be limited to *MEDLINE* or non-*MEDLINE* records. CAS Registry Numbers are available in many records to identify chemical substances. Chemical terms are indexed so that they can be searched as segments within other chemical terms or as the complete term.

TOXLINE is composed of the following subfiles: *Aneuploidy File*, pre-1950 to the present; *Environmental Mutagen Information Center File (EMIC)*, 1950 to the present; *Environmental Teratology Information Center File (ETIC)*, 1950 to the present; *Epidemiology Information System*, 1940 to the present; *Federal Research in Progress*, 1982 to the present; *Hazardous Materials Technical center*, 1982 to the present; *International Labour Office, CIS Abstracts*, 1981 to the present; *International Pharmaceutical Abstracts*, 1969 to the present; *NIOSHTIC*, 1984 to the present; *Pesticides Abstracts*, 1968-1981; *Poisonous Plants Bibliography*, pre-1976; *Toxic Substances Control Act Test Submissions*, all relevant submissions; *Toxicity Bibliography*, 1965 to the present; *Toxicological Aspects of Environmental Health*, 1970 to the present; *Toxicology Document and Data Depository File*, 1979 to the present; *Toxicology Research Projects, National Institutes of Health (CRISP)*, 1986 to the present. Additionally, DIALOG has included toxicology-related records from *International Pharmaceutical Abstracts* and the BIOSIS subfile *Toxicological Aspects of Environmental Health*.

The *TOXLINE* database (DIALOG File 156) consists of 1.5 million records; it is updated monthly with 10,000 records. The cost for searching the file is $1.10 per connect minute. Print charges are $.20 per full record printed online or offline.

TSCA Chemical Substances Inventory Reloaded

DIALOG's File 52, *TSCA Chemical Substances Inventory* has been reloaded and now contains the 1990 supplement to the 1985 edition of the *Toxic Substances Control Act (TSCA) Inventory*. The file is a compilation of information on chemical substances manufactured, imported or processed for commercial purposes in the United States. The *Inventory* was developed in response to Section 8(d) of the Toxic Substances Control Act, U.S. Public Law 94-469. The database includes all substances listed in the 1979 *Initial Inventory* the 1980 *Supplement*, the 1983 *Cumulative Supplement*, the reissued *Inventory* of May 1985, and the March 30, 1990 reissue. The cumulative nature of the file makes it a valuable historical resource in addition to its usefulness as a list of chemical substances contained in the current *TSCA Inventory*. Each record provides the following information for each substance listed: CAS Registry Number, preferred name, synonyms, molecular formula, and date(s) of publication in the *Inventory*.

Federal Acquisition Regulations Online

A new file containing the full text of the *Federal Acquisition regulation (FAR)* and its supplements is now on DIALOG. *FAR* is the United States government publication that dictates the implementation of U.S. government procurement policy. It outlines the regulations federal executive agencies must follow in the acquisition of supplies and services. *FAR* is issued and maintained by the Department of Defense, the General Services Administration, and the National Aeronautics and Space Administration. Coverage includes regulations pertaining to more than twenty-five federal agencies, including the Department of Health and Human Services, the Department of Defense, the Environmental Protection Agency, the Department of Labor, and the Department of Energy. The file is kept current with the addition of material from federal Acquisition Circulars and Defense Acquisition Circulars. Notices and Proposed Rules and Regulations appear first in the *Federal Register*.

The price for searching *Federal Acquisition Regulations*, DIALOG's File 665, is $2. per connect minute. Full records can be printed online or offline for $.50 each. The database covers 1984 to the present, and contains 32,000 records. Twenty-five new records are added with each weekly update.

Wilson Databases on BRS

Maxwell Online has announced the addition of the seventeen databases produced by the H.W. Wilson Company. They will be added to the BRS Search Service throughout 1991. The first database to be brought up is the *Readers' Guide to Periodical Literature*. Of special interest to sci-tech libraries are the following: *Applied Science & Technology Index*, *Biological & Agricultural Index*, and *General Science Index*. Wilson's other databases, containing general business, legal, and book information could also prove useful. Search costs have not yet been determined. Wilson is also offering the databases through Maxwell's BRS/OnSite program.

SEARCH SYSTEM NEWS

DIALOG Menus Announced

DIALOG has announced a new mode of searching, *DIALOG Menus*, intended for the infrequent searcher and/or the end-user. It provides access to more than 220 of the most popular DIALOG databases. Everyone with a DIALOG password has access to this feature. From the *Menus* main menu, one can choose from broad subject categories, including agriculture, food, and nutrition; biosciences and biotechnology; chemistry; company information; computers and software; energy and environment; engineering; industry analysis; medicine; and patents. Searches then proceed to choose subcategories. Once a database is selected, the searcher is guided to select the type of search, and then to enter search terms. *Menus* also facilitates display of search results, transferring to another database, and multiple file searching. DIALOG states that this search interface has been used in their Corporate Connection program for the last two years.

A free copy of *Searching DIALOG: The Guide to Menus* was distributed to all DIALOG customers. This 96-page manual includes a tutorial section, search tips, telecommunications information, a glossary, and an alphabetical list of databases with brief descriptions, acronyms, and DIALOG file numbers. Copies of the manual may be ordered for $5.

ELHILL Enhancements Added

The National Library of Medicine (NLM) has provided several new features to make the retrieval of data from the ELHILL databases easier and more efficient. New search capabilities include text word searching of the title field and the address field; addition of two new fields (publication type and gene symbol); pre-explosions of "adult" and "child" which permits retrieval of all age groups subsumed under each of these terms in one step; and subheading pre-explosions.

The addition of the gene symbol to *MEDLINE* and related

ELHILL databases represents a further cooperative effort between NLM and the National Center for Biotechnology Information. This new field will contain the "symbol" or abbreviated form of gene names as reported in the literature. The gene symbol field will represent the first use in ELHILL databases of Standard Generalized Markup Language, a relatively new manuscript markup language for electronic publishing.

New Files from the American Petroleum Institute

APIPAT (Files 353, 953) and *APILIT* (Files 354, 954) are now available on DIALOG. Produced by the American Petroleum Institute's (API) Central Indexing & Abstracting Service, these files are designed to serve the needs of the petroleum and petrochemical industries. Both databases include coverage since 1964. *APIPAT* contains citations to worldwide patents; before 1982, coverage was limited to patents issued in nine major countries and by the World Intellectual Property Organization. *APILIT* provides worldwide coverage of non-patent literature pertaining to the petroleum and petrochemical industries. Records are drawn from a variety of publications, including trade and technical journals, government reports, meeting papers, and API publications.

APIPAT includes patents relating to petroleum refining and petrochemical technology' fuels, lubricants, and other petroleum products, catalysts, additives, pollution control, waste disposal, transport and storage, synthetic fuels, and oil filed chemicals. *APILIT* subject coverage includes petroleum processes, petrochemicals, fuels, alternate energy sources, lubricants, pipelines, and health, safety, and environmental issues. Oil field chemicals information has been included in both databases since 1981.

Files 353 and 354 are open to non-API subscribers; Files 953 and 954 are available only to subscribers. Non-subscribers are limited to a maximum of combined usage of three hours per calendar year in the files. Abstracts are searchable but not displayable in the open files. API subscribers have unlimited use of Files 953 and 954, which offer discounted pricing. Costs for searching *the databases* are $3.08 per connect minute (File 353, 354) and $1.83 per minute (File 953, 954). Prices for printing records range form $.35 to $.60. *APIPAT* contains 185,000 records; it is updated monthly with 650 new records. *APILIT* contains 420,000 records, with 1300 new records added each month.

PUBLICATIONS AND SEARCH AIDS

PsycINFO Publishes New Thesaurus

The sixth edition of PsycINFO's *Thesaurus of Psychological Index Terms* is now available. This reference tool has been revised, updated, and expanded to increase the vocabulary's accuracy, completeness and usefulness. The thesaurus includes all terms used to index documents in the database. It is also used to compile the subject indexes for the printed *Psychological Abstracts* and *Psyc-BOOKS*. More than 230 new index terms, 100 cross references and 120 new scope notes have been added to this addition. Revised hierarchies and updated postings notes are also included. A new feature, index term clusters, was developed for users unfamiliar with the thesaurus vocabulary. It groups approximately 2600 index terms into eight broad categories and subcategories. The thesaurus is priced at $65 list, or $49 for members of the American Psychological Association (APA). The 1991 *Thesaurus of Psychological Index Terms* may be ordered from the APA Order Department. For further information, contact PsycINFO User Services, 1400 North Uhle Street, Arlington, VA 22201, (800) 336-4980 or (703) 247-8929.

EDUCATION

BRS Forms Pharmaceutical Focus Group

In an effort to better serve the needs of the information scientist in the pharmaceutical industry, BRS has created a Pharmaceutical Focus Group. This half-day session is designed to keep searchers up to date about all new BRS features and databases. Sessions include searching techniques, tips for the end user, new pharmaceutical databases, and information on BRS's business databases. At least four such focus groups are scheduled for the year. For more information, contact Phil Adams, BRS Pharmaceutical Sales Representative at (301) 621-7590.

For Product Safety Concerns and Information please contact our EU
representative GPSR@taylorandfrancis.com
Taylor & Francis Verlag GmbH, Kaufingerstraße 24, 80331 München, Germany

www.ingramcontent.com/pod-product-compliance
Lightning Source LLC
Chambersburg PA
CBHW061156220326
41599CB00025B/4501